JAGUAR

The iconic models that define the marque

JAGUAR
3.8 LITRE

JAGUAR

The iconic models that define the marque

Colin Salter

Editor
Paul Walton

PAVILION

CONTENTS

INTRODUCTION: THE JAGUAR STORY

The history of Jaguar is the history of the British motor industry itself. It's about humble origins and British ingenuity; it is about global success and cars that would become famous all over the world. But it is also about decline and the eventual takeover by a foreign entity.

At the beginning, though, it was just about one man. Born in Blackpool on 4th September 1901, William Lyons was the son of William, an Irish musician turned piano tuner, and Mary Jane Barcroft. The young Lyons attended the local grammar school and wasn't a keen student, however he did have an affinity with all things mechanical. He bought his first motorbike when he was a teenager, a 1911 Triumph, the first of many that would be parked outside the family home at King Edward Avenue.

Living a few doors down from the Lyons family home was William Walmsley. A former solder, nine years old than Lyons, Walmsley was another motorcycle enthusiast and had designed a handsome bullet-shaped sidecar, its aluminium bodywork left unpainted. Other local bikers became interested in the design and he began producing the sidecar in his garage on a limited basis. The sidecar soon caught the attention of Lyons who subsequently bought one. The two Williams struck up a friendship and in 1922 went into business to build and sell the sidecar on a larger scale.

The newly named Swallow sidecar was a success, but the ever-ambitious Lyons wanted more and in 1927 the company began offering a coachbuilt version of the Austin Seven. Knowing Blackpool didn't have the car-producing skills Lyons required if he was to grow his

Left: The car that started it all, the SS1.

Below: Known as the 'Leaper' Jaguar's first mascot was described by Sir William Lyons as "looking like a cat shot off a fence". The second, by *Autocar* illustrator F. Gordon Crosby, has stood the test of time.

Above: Two of the five D-types entered in the 1957 Le Mans. The No.17 car of Lucas/Brousselet would finish third and the No.16 car of Frère and Rouselle would come fourth. It was the year Jaguar finished 1, 2, 3, 4, 6.

Above left: *Motor* magazine from September 1956 advertising the 2.4-litre saloon, which would subsequently be known as the Mark 1.

Left: Launched in 1948 the XK120 was the sensation of the London Motor Show.

fledgling company, he moved it to Coventry, the heart of British car building. Other coachbuilt models soon followed based on Fiat, Standard, Swift and Wolseley chassis. The turning point in the company's history came in 1930 when Lyons came to an agreement with the Standard Motor Company for it to provide chassis and engines to be fitted with Lyons' own stylish bodies. The first was the handsome SS1. Its long bonnet, low roofline and rakish good looks would become a Lyons' trademark over the coming decades.

The Jaguar name first arrived in 1935 manifesting as a range of cars – the SS Jaguars – that included a four-door saloon and a beautiful two-seater sports car. Featuring a new 2.7-litre six engine, the SS90 was as fast as it was handsome, as was the SS100 that followed and featured an upgraded 3.5-litre version. This was Lyons' first true sports car and every Jaguar model since can trace its lineage directly to this car.

With Walmsley leaving the company in 1935, by the time war erupted four years later Lyons had transformed his humble sidecar company into a serious automobile manufacturer, exporting his cars to the lucrative American market. But the company's engines were still produced by Standard and whilst fire-watching at the height of the war from the factory roof, engineers Claude Bailey, William Heynes and Wally Hassan dreamt up an engine that would transform the company's cars into a class of their own.

After the hostilities had ended (and the company had dropped its now politically awkward SS name to plain Jaguar) production of the company's post-war models was restarted. Its first all-new model, the Mark V, was a handsome saloon, but very much a stopgap. Behind the scenes a brand new car was being developed that would feature the new engine dreamt up on the factory roof that would replace the old pushrod motor.

Wanting to publicise this new XK engine unit before the saloon was ready (it wouldn't appear until 1950), at the 1948 London Motor Show Jaguar unveiled a stunning new sports car. Using a cut-down chassis from the forthcoming saloon, it featured a 3.4-litre version of the new engine (a 2.0-litre also planned, but never realised). With its beautiful voluptuous body and claimed top speed of 120mph (a huge figure for the era) the XK120 was the star of the show.

It wasn't long before the car reached the track, with its first victory being the 1949 Daily Express One Hour Production Race at Silverstone. By the following year, XK120s were entered in all of the major European races, including the Targa Florio in

Sicily, France's 24-hour race at Le Mans and the epic Mille Miglia, the 1000-mile dash from Brescia to Rome and back again. In September, a young Stirling Moss took the car's first major victory at the Dundrod TT in Northern Ireland. To begin with Lyons had no interest in racing; the company was more concerned with getting the new saloon, the Mark VII, (there was no Mark VI) into production. But eventually he could see the benefits of racing and after fielding a semi-works team of XK120s for both rally and track events (the most famous being the Alpine Rally-winning car, registered NUB 120), the car was transformed into a genuine race car. In 1951 Jaguar employed a former aeronautical aerodynamicist, Malcolm Sayer, to do it. Although the car had a new tubular chassis frame and a more aerodynamic body, it was christened the XK120C, which was soon shortened to just C-type. With the 3.4-litre XK engine under the bonnet, the car was quick and reliable, as proven when it won the 1951 Le Mans, a feat it repeated two years later.

Jaguar's range was slowly developed, with the XK120 giving way to the XK140 in 1953 that featured bigger bumpers and more precise rack and pinion steering, but Jaguar's biggest step forward was with the 2.4 Mark 1 saloon from 1955. As the first Jaguar to feature a monocoque chassis the Mark 1 was a small and nimble saloon that, when joined by the 3.4 version, was fast too.

In 1954 the company had revealed the Malcolm Sayer-designed D-type, a brand new purpose-built racing car that featured improved aerodynamics and, from 1955, a 3.8-litre version of the XK engine. As the car to beat during the mid to late 1950s, the D-type won Le Mans an impressive three times in 1955, 1956 and 1957. Jaguar pulled out of racing at the end of 1956, their official works car having been beaten that year by the talents of privateer team Ecurie Ecosse. David Murray's team bearing the Scottish Saltire would go on to lift the trophy again in 1957.

By the end of the 1950s, the car had lost its competitive edge, not helped by new rules in sports car racing which outlawed engines over 3.0 litres. With Jaguar still having unsold examples, the factory began to convert these into a road-going sports car called the XKSS. The argument as to whether this was a stop-gap solution to use up expensive racing equipment or a long-term plan was cut short in February 1957 when a devastating fire raged through Jaguar's factory at

Above: The wreckage of the Browns Lane factory in February 1957. Charred bodyshells of XK140 cars are suspended over burnt-out Mark VII saloons.

Browns Lane in Coventry (the former shadow factory the company had moved into during 1951). Nine of the 25 cars that were in various stages of completion were destroyed, as were the car's jigs and tooling, effectively ending production.

Although the XKSS was very much Jaguar's past: 1960 saw the first hint of the company's future. At that year's Le Mans the American racing team, Briggs Cunningham, entered a new type of Jaguar. With its cowled lights and oval grille, it was clearly inspired by the D-type but was larger and with better proportions. This was the E2A, a one-off prototype that didn't even finish the race yet is hugely significant as the forerunner of the car that would come to define Jaguar and the decade it came from.

Unveiled at the Geneva Motor Show in 1961, the E-type stunned the world with its amazing looks. Enzo Ferrari called it, "The most beautiful car ever made". Designed again by Malcolm Sayer, it was the result of everything he'd learnt about aerodynamics during the company's racing years, while under the skin it featured a brand new independent rear suspension layout plus the faithful 3.8-litre XK unit. This layout was also used for Jaguar's new large saloon, the Mark X which, like the Mark 2 from 1959, had a unitary chassis.

As the 1960s unfolded, the E-type was constantly upgraded and modified; the 4.2-litre version of the XK engine arrived in 1965, while a stretched 2+2 coupé was added to the range in 1966. Although not as pretty as its two-seater sibling, the 2+2 was important because it made the car slightly more practical for families. A lightweight competition model was also developed by the firm's competition department and was campaigned successfully by several private teams.

Jaguar came close to officially returning to competition in the mid-1960s with the Sayer-designed XJ13. A handsome, mid-engined sports car that was powered by a 5.0-litre V12, its development had been part-time and ultimately took too long. By the time the sole car was finished in 1966, its design had already become obsolete, its engine too large and the project wasn't taken further.

By 1966 Lyons was in his mid-sixties. With no successor following the death of his only son John, who had been killed in a car accident en route to Le Mans, he was looking for future stability and so joined BMC (British Motor Corporation – an agglomeration of the Austin and Morris car firms along with MG, Riley and Wolseley) and Pressed Steel – the company that made the bodies for Jaguar's cars – to form British Motor Holdings. Sir William retained the title of Jaguar's chairman and the company remained largely autonomous. But BMH didn't last long and in 1968 they amalgamated with Leyland, the bus and truck manufacturer, which had recently acquired Standard-Triumph and Rover. This resulted in the formation of the British Leyland Corporation, a huge conglomerate of eleven car marques, of which Jaguar was now just a tiny part.

The E-type Series 2 arrived the same year and featured a larger grille, open headlights, full wrap-

Left: Two advertisements from Jaguar emphasizing the company's favourite slogan, 'Grace, Space, Pace'.

Below: The launch of the Jaguar E-type, known as the Jaguar XKE in the United States, confirmed Malcolm Sayer's reputation as a car designer of genius.

around bumpers plus larger rear lamps that were now situated below the bumpers. Inside, the dashboard was also redesigned with new rocker switches that met U.S. safety regulations. Cooling was improved thanks to a larger radiator and twin electric cooling fans, plus the car's braking was substantially uprated.

Also making its debut in 1968 was Jaguar's groundbreaking saloon, the XJ6. In one swoop the car replaced the company's complicated saloon line-up that consisted of the 420G (a 4.2-litre-engined Mark X), the S-type (a stretched Mark 2 with a boot) and the 420 (an S-type with a similar nose to the Mark X's with four headlights) plus, following Jaguar's acquisition of the company in 1960, a Daimler-badged version of the latter called the Sovereign.

Behind the scenes Jaguar had been working on its first new engine since the XK unit from 1948, a 5.3-litre V12. Aimed squarely at the American market, it would become a Jaguar mainstay for the next 30 years. The first car to feature the engine was the E-type Series 3 from 1971, making it a very different car from the earlier models. The V12 version of the XJ arrived in 1973.

Thanks to proposed (but never passed) American legislation that would have outlawed convertibles, the E-type's replacement took Jaguar in a new direction. Large, luxurious and a 2+2 coupé only, the 1975 XJ-S wasn't immediately popular and with only one engine option, the V12, poor economy didn't help its cause either. Sales were so bad that by 1980, when just 1,057 were sold, production ceased for a short time.

Help arrived in the shape of John Egan who became Jaguar's chairman in 1980. This former Massey Ferguson director soon instigated a comprehensive drive to improve the build quality of Jaguar's cars which had slumped under the management of British Leyland. The problem of Jaguar's economy was also partially solved thanks to a redesigned cylinder head for the V12, plus an all-new 3.6-litre straight-six engine. Knowing the company needed a halo model, Egan

Below: A brochure image of the Jaguar XJ-S range of 1984, with coupé and cabriolet but still awaiting the factory convertible.

was behind a cabriolet based on the XJ-S from 1983 (a very handsome full convertible would arrive five years later). The difference these changes made to the car's sales was immediate – in 1984 6,028 XJ-Ss were produced. The company would become independent again in July 1984.

Sales were so promising that in 1982 Jaguar went back to racing for the first time since the 1950s when it entered the European Touring Car Championship (ETCC). This wasn't the first time the XJ-S had headed to the track; the American Group 44 team had enjoyed some success with the car in the late 1970s. But the European Touring Car Championship effort, masterminded by Tom Walkinshaw Racing (TWR), was more significant. Not only was Walkinshaw himself crowned the 1984 ETCC champion but the campaign would lead Jaguar back to sports car racing and two further victories at Le Mans in 1988 and 1990.

The most important new saloon during the Eighties was the XJ40-generation of XJ that replaced the XJ Series 3 in 1986. A brand new car that was partly designed by Sir William Lyons (who died in February 1985), it was the first Jaguar to feature onboard diagnostic and engine management systems. Pushed through development too quickly, the car wasn't ready for production and early models suffered from poor reliability. Although this was slowly improved, the car would tarnish Jaguar's image for years.

Yet by the end of the 1980s Jaguar's fortunes had improved to such a degree it was the ideal target to be taken over. After coming close to a deal with General Motors, Jaguar was eventually bought by Ford for £1.6 billion in 1989.

The first car to be launched under Ford's ownership was a special one, the XJ220. Starting life as a concept in 1988, it was subsequently put into production by a

Above left: Winning at Le Mans in the 1980s proved to be more difficult than in the 1950s, but in 1988 Jaguar won the great race again, with the No.2 TWR Jaguar XJR-9LM of Jan Lammers, Johnny Dumfries and Andy Wallace.

Left: Keen to emphasize their sporting heritage, new Jaguar models are often posed with victorious sports cars at Goodwood, or in this case, the old start/finish straight at Silverstone. Sharing the tarmac with a D-type is the XK180 concept car designed by Keith Helfet.

satellite operation that was owned jointly by TWR and Jaguar. With a top speed of 217mph this was a fully paid up member of the supercar club, a rival to the Ferrari F40 and Porsche 959.

The XJ-S was given a £50m facelift in 1991, but the most significant car of the 1990s was the XK8 that replaced the XJ-S in 1996. With soft curves that were clearly inspired by the E-type, this handsome car was very different from the angular XJ-S. More importantly, power came from a brand new 270bhp 4.0-litre V8 that replaced the straight six and the V12 that by now had been enlarged to 6.0 litres. A V8-engined XJ saloon (codenamed the X308) soon followed and both cars had a supercharged version that with 370bhp made them the fastest mainstream production cars in Jaguar's history.

A one-off car based around the XK8, the XK180, was created to celebrate the XK engine's 50th anniversary in 1998. Small, handsome and reminiscent of the D-type, the car was well received and Jaguar designers looked at creating a production version. Sadly the F-type Concept never got further than a single prototype which was unveiled at the 2000 Detroit Motor Show. Purists would have to wait a little longer for a genuine two-seater Jaguar sports car.

With Ford money now being pumped into the company, Jaguar could branch out into other new markets. The first was the retro-styled 1998 S-TYPE (named and designed after the 1960s saloon of the same name) that would compete with the BMW 5-Series in the executive market. This was followed by the smaller Ford Mondeo-based X-TYPE a couple of years later. A new XJ arrived in 2003 that, although still following the XJ's traditional design, was constructed from aluminium, producing a car that was both light and rigid. The majority of Jaguar's future models would be produced from the material.

This included the XK8/XKR's replacement in 2005, the new XK, which had been developed under the nametag Advanced Lightweight Coupé Concept. The first car to be entirely designed by new studio boss, ex-Aston Martin man Ian Callum, it was a handsome

Above: Jaguar's most innovative concept car yet, the C-X75 was launched at the 2010 Paris Motor Show.

Right: The Jaguar XKR-S, XJ and XFR line up in front of Goodwood House and a modular sculpture of an E-type, built to celebrate Jaguar at the 2011 Goodwood Festival of Speed.

coupé and convertible that again was powered by the V8, although now enlarged to 4.2 litres.

Sadly, despite Jaguar's very expensive 2000-2004 Formula One effort that raised the company's profile but saw little in the way of results, sales were poorer than expected. After an international high of 130,322 in 2002, that had slipped to 120,000 a year later. Ford tried to save costs, including the closure of Jaguar's famous Browns Lane assembly plant in 2005. But the writing was on the wall. In 2007 Ford made it known the company was for sale and after being courted by several suitors, Jaguar was eventually sold (along with Land Rover to create a new joint company, Jaguar Land Rover) to the Indian industrial giant, TATA.

With a very enthusiastic new owner, Jaguar's rebirth could begin. The XF replaced the S-TYPE in 2008. It may have used the old car's chassis but it had a fresh and modern look that pointed towards Jaguar's future. An estate version, called the Sportbrake, arrived in 2012. An Ian Callum-designed XJ luxury saloon, codenamed the X351, was revealed in 2009. It followed similar styling cues from the XF, making it the biggest departure from the model's traditional design since the XJ40.

In 2010 Jaguar enlarged the capacity of its V8 to 5.0 litres, which when supercharged to over 500bhp, boosted the performance of any car it was fitted to. The XKR became a harder-edged sports car, with the ultimate version being the 2013 XKR-S GT. With power increased to 542bhp and lowered, lightened and coming with a full aero package, the car was the clearest illustration yet of Jaguar's new found confidence.

This was proven again in 2013 when the first proper Jaguar sports car since the end of the E-type 40 years earlier was revealed. Again produced from aluminium, the F-TYPE was a modern car but with a few subtle nods to its famous forebear. Power came from a 3.0-litre V6 or the 5.0-litre V8 and it replaced the XK/XKR coupé.

With new cars coming thick and fast during the second half of this decade – including Jaguar's first SUV, the F-PACE, the ultimate track-day car the F-TYPE Project 7, and a whole new Special Vehicles Division – Jaguar's future at the forefront of automotive technology and design looks as exciting as it is assured.

Paul Walton
Editor, *Jaguar World*

SS1

There are many important cars in the history of Jaguar; the XK120, the D-type, the Mark 2, E-type, XJ6 and XK8, but none is as important as the SS1. Its launch was a declaration of intent, an intimation of potential, and if it hadn't worked there might never have been a Jaguar car company.

On his twenty-first birthday William Lyons (1901-1985) founded the Swallow Sidecar Company with his business partner William Walmsley. Walmsley already ran a small but successful business from the garage of

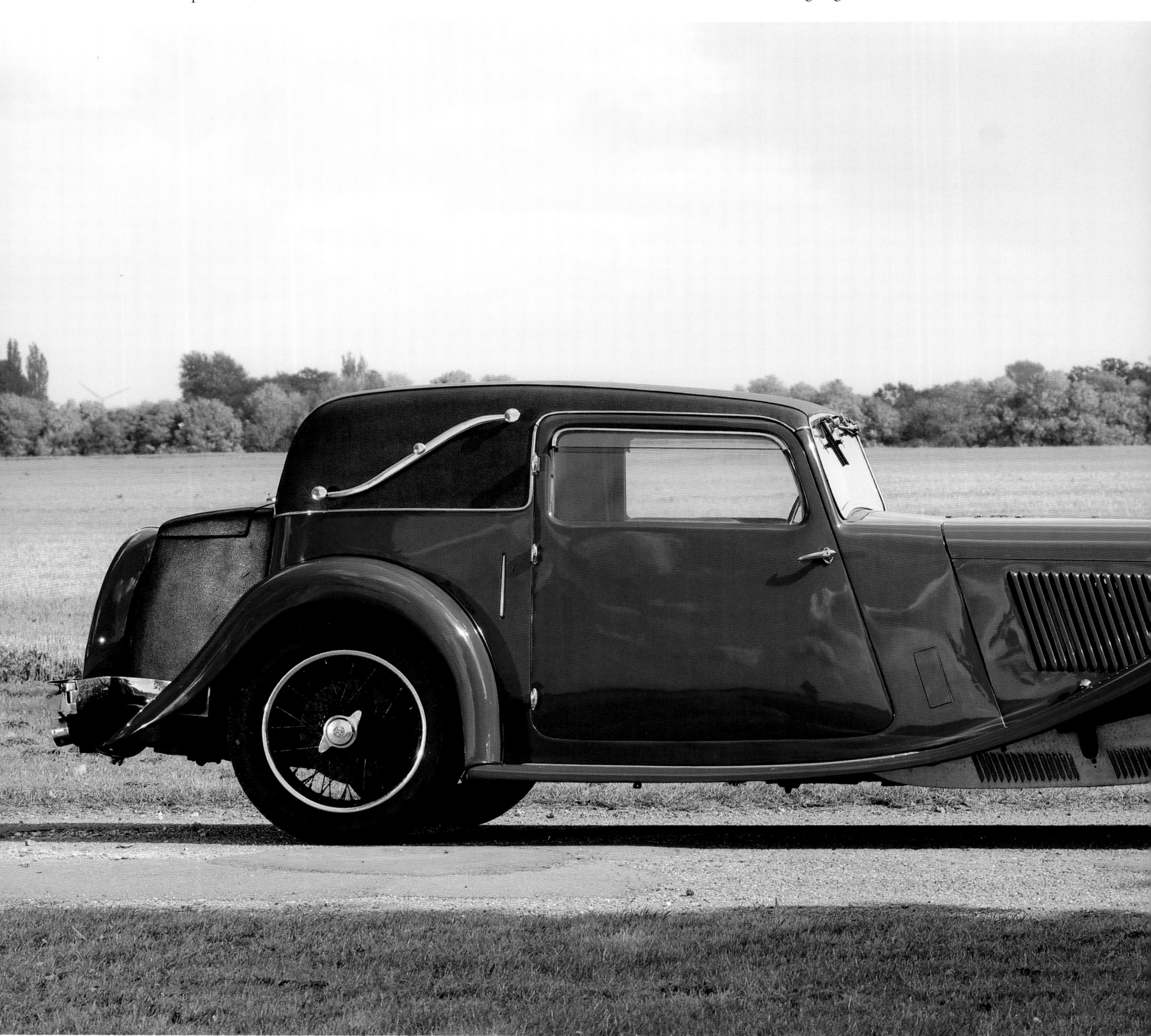

his parents' home in Blackpool, Lancashire; where he built stylish, bullet-shaped sidecars and attached them to army-surplus motorcycles which he sold to civilians. Lyons had bought one a year earlier, and now his business acumen and eye for style saw Swallow Sidecars prosper and expand from its original workshops in Blackpool's Bloomfield Road to new ones in Cocker Street.

In 1927 their success and ambition encouraged Lyons and Walmsley to build their first four-wheeler. Based on the Austin Seven, it was christened the Swallow. At the same time the business dropped the word Sidecar from its name and became the rather grander Swallow Coachbuilding Company.

The Austin Seven was the most popular British car at the time; the Model T Ford of its day, but only half the size. It was a basic car for the mass market; a 747cc side-valve, four-cylinder engine, with three gears and no frills in terms of comfort and style. Lyons asked Swallow's coachbuilder Cyril Holland to design a new body to mount on an Austin Seven chassis. Holland delivered the Swallow, with smart new curves – a round-nose radiator and a plump, rounded rear end compared to the original Seven's boxy look.

Although the Swallow's beating heart was still just the Austin Seven's small engine, its elegant shape and glamorous two-tone colour schemes captured the public imagination at a time of economic austerity. When Henly's, a large London car dealer, placed an order for 500 vehicles with the company, Lyons knew he had a success on his hands. To fulfil the Henly's order the company had to expand rapidly. Blackpool, the archetypal British seaside holiday resort, simply

Left and Below: The original model of the SS1 was launched with a less elegant 'helmet wing'. This 1932 revised model SS1 can be distinguished by the continuous wingline flowing from front to rear.

SS1

ENGINE inline six-cylinder

CAPACITY 2663cc

BORE X STROKE 73 x 106mm

COMPRESSION RATIO 7:1

POWER 62bhp

Valve gear side-valve

FUEL SYSTEM Zenith carburettor

TRANSMISSION 4-speed manual

FRONT SUSPENSION half elliptic transverse spring

REAR SUSPENSION cantilevered quarter elliptic springs

BRAKES drum, front by handbrake, rear by foot pedal

WHEELS wire, bolt-on

WEIGHT 2967 lbs (1346 kg)

MAXIMUM SPEED 75mph (121kmh)

PRODUCTION 4254, 1931-1936

did not have the industrial base from which to supply a suitably skilled workforce, and so in 1928 Lyons moved the entire Swallow operation to Coventry, then as now the centre of the British motor industry.

The Austin Seven Swallow was still essentially an Austin Seven, and it was only a matter of time before Lyons determined to build a car to his own specification. Although not yet ready to construct his own chassis, he commissioned the Standard Motor Company to produce a chassis to suit his requirements. In 1931 Swallow unveiled the SS1 at the London Motor Show, using the new Standard chassis and a Standard engine but with a stylish body that was all Swallow's own work.

Lyons sketched out the original design himself, including the strikingly sporty stretched bonnet. He and Walmsley disagreed about the height of the roof and the clearance of the car – Walmsley wanted both to be higher – and when Lyons was taken ill during the design process, Walmsley pushed his ideas through. Although Lyons didn't like them, they were in tune with contemporary fashion and the launch was a success. But as production numbers passed the 500 mark, the chassis of the SS1 switched from overslung to underhung, giving it the lower clearance which Lyons had wanted. The move emphasised the length of the car and improved its handling on the road.

The SS1 was eventually available as a tourer, a coupé and as a saloon. Although it was promoted as 'the car with the £1000 look', the SS1 carried a price tag of only £310. The interior styling matched the wealthy appearance of the exterior – rich leather upholstery, a wooden dashboard and wood-panelled doors with a tooled leather art deco, rising sun motif. This affordable luxury was enough to overcome, in the public's mind, the car's rather underwhelming performance. The long bonnet covered an engine short on power and speed. Standard had provided them with a six-cylinder side-valve engine with a capacity of 2054cc producing 48bhp, and from 1932 to 1934, a 2552cc engine generating 62bhp. For the 1934 to 1936 models engine size was increased to 2143cc producing 53bhp or alternatively 2663cc giving 68bhp. In 1935 Lyons lowered the high roof which Walmsley had imposed on the saloon, by introducing a new Airline coupé version of the SS1 with a sculpted rear end.

It was with the SS1 that Lyons made his first real mark in the world of cars. His sense of style, his

Above left: An SS1 2+2 coupé with rear window. Swallow also produced a pillarless 'Airline' coupé in 1935 with a larger rear window and curved 'fast back'.

insistence on top quality craftsmanship, and his willingness to address problems, to adapt his ideas to achieve greater success, all demonstrated his ability to take a place among the major players in the industry.

In any case, the SS1 was joined in 1932 by the SS2, a shorter version of its predecessor, selling at £210. Although vulnerable to the same accusations of a lack of power, it also shared the SS1's sense of graceful, racy style. Like the SS1, the SS2 received an engine upgrade in 1934, and was available in coupé, tourer and saloon versions.

The success of both models persuaded Swallow to cease production of the old Austin Seven Swallow. And from 1934 onwards the company, formerly known as Swallow Sidecars, underwent yet another change of name and became SS Cars Ltd. Although motor cars were now its principle business, it continued to manufacture sidecars well into the 1940s. But William Lyons' co-founder William Walmsley had come to the end of his road in the partnership.

Injured in the Great War and far less ambitious than Lyons, Walmsley was happy with the company's success to date but not interested in further growth. It is claimed that he spent more time in the company workshops making parts for his model railway than in developing the company's products. Lyons bought him out in 1935, and Walmsley invested his time and money in the design and manufacture of caravans and trailers. From now on until his retirement in 1967, William Lyons was the sole Managing Director of the company soon to be known as Jaguar.

Above: An extension in wheelbase in 1932 allowed more legroom in the back and the installation of armchair-like rear seats.

Right: An open tourer was added to the SS1 range in March 1933.

SS JAGUAR 100

The SS100 was a product of its time, the age of speed; the same half-decade saw the production of the streamlined Class A4 steam engines *Sir Nigel Gresley* and *Mallard*, and the Supermarine Spitfire aircraft. The first new car to emerge from the SS Cars production line after William Walmsley's departure was the SS90, a lovely-looking open-top sports car which drew its name from its optimistic claim to achieve 90mph.

The number was stamped on a plate on the headlamp mount at the front of the car – rather rashly, because the SS90 soon attracted the same criticisms levelled at the SS1 and SS2, namely that it was "more show than go".

Although styled as a sporty two-seater tourer, the SS90 was in essence an SS1 cut down and mounted on an SS2 wheelbase. The 2.5-litre, side-valve, six-cylinder engine of the SS1 left the SS90 as underpowered as the earlier vehicle had been. The market was unimpressed, and only 23 of the SS90 were ever produced.

William Lyons called in cylinder-head specialist Harry Weslake to address the problem. His brief was simple: raise the Standard SS1 engine's output from 70 to 90bhp. At the same time, Lyons appointed a new chief engineer at SS Cars, Bill Heynes. Heynes had been head of the technical department at Humber Cars in Coventry, for whom he had worked since 1922 – the same year Lyons teamed up with William Walmsley. Heynes' early contributions at SS were to chassis design, but he also turned his attention to the output of the Standard engines, and became a key player in the Jaguar story for over 30 years.

Between them they more than answered Lyons' brief: Heynes reworked the old SS1 Standard Motor Company chassis, strengthening it and improving the braking systems. Weslake's new overhead valve design delivered a transformative 102bhp. Standard manufactured the new head for SS Cars, who would use it on the SS Jaguar 100, one of a range of new cars introduced in the mid-1930s comprising the SS Jaguar 1½ litre, SS Jaguar 2½ litre, SS Jaguar 3½ litre and the SS Jaguar 100.

The new car was launched at the London Motor Show of 1936 and for the first time in the company's history, the model had been given a name as well as a number. William Lyons had a friend in the Royal Flying Corps (the predecessor of the Royal Air Force), a mechanic who had spoken enthusiastically to Lyons about one of the aircraft engines on which he worked in the 1920s: the Armstrong Siddeley Jaguar. The then unfamiliar word struck Lyons, and now it seemed the perfect fit for his new vehicle. Although not present on the very earliest models, the leaping Jaguar mascot which still graces Jaguar cars today, made its first

Left: The SS Jaguar 100 was preceded by the SS 90. Both cars carried the type name in the centre of the lamp bar.

SS JAGUAR 100

ENGINE inline six-cylinder

CAPACITY 3485cc

BORE X STROKE 82 x 110mm

COMPRESSION RATIO 7.2:1

POWER 125bhp

VALVE GEAR overhead camshaft

FUEL SYSTEM twin SU carburettors

TRANSMISSION four-speed manual, rear-wheel drive

FRONT SUSPENSION solid axle, semi-elliptic leaf springs, hydraulic and friction dampers

REAR SUSPENSION underslung live axle, semi-elliptic leaf springs, hydraulic dampers

BRAKES Girling rod-operated drums

WHEELS wire, bolt-on

WEIGHT 2609 lbs (1183 kg)

MAXIMUM SPEED 104mph (167kmh)

PRODUCTION 314, 1936-1940

FPD 634

Top: The first in the SS Jaguar range, the Jaguar 2½ litre, was originally presented as a four-door saloon. When it was launched it drew favourable comparisons with similar Bentley models.

appearance on the SS Jaguar.

The SS Jaguar 2½ litre was originally presented as a four-door saloon (perhaps to distance it from the unsuccessful open-top SS90). Dealers and press were invited to view it at a lunch in the Mayfair Hotel in London, and asked to write down their estimates of its selling price. The average guess was £765. In fact its combination of style, power and luxury would cost a mere £395 at a time when a new Rolls Royce cost in the region of £1500. A pared-down 1½ litre was also available, and within a year an even more powerful, 3½ litre option was introduced with a £450 price tag. Its nearest rival was the Morris MG SA, which was launched the same year with a mere 2.0-litre engine.

It was only a matter of time before the Jaguar appeared in its most familiar form, a two-seater sports car, the SS Jaguar 100. The exaggerated curves of the sweeping wings attracted some criticism from those seeking aerodynamic perfection; and others frowned on the foldaway windscreen and cutaway doors. But for almost everyone else, this version of the SS100 was and remains the epitome of pre-war sports car motoring.

Achieving speeds of over 100mph (in its 3.5-litre form), there was nothing else like it on the roads of Britain; but its performance was turning heads further afield too. British motorsports journalist Tommy Wisdom and his wife Elsie surprised the motoring world when they completed without penalty the Eighth International Alpine Trial in 1936. Wisdom, driving an unmodified 2.5-litre works version of the car, took a very creditable second place in a one-kilometre time trial at St Moritz. In 1937 he drove the fastest lap at Brooklands in Surrey, pushing the same car to 118mph.

Grand Prix driver F.J. McEvoy won his class on the track at Marnes near Rheims in 1936, and in 1937

Portuguese racing driver Casimiro de Oliviera drove an SS Jaguar 100 to overall victory in front of a home crowd in a race on the Circuit International de Villa Real. It was the Jaguar's first outright win on the continent, achieved over Alfa Romeo, Aston Martin, BMW and others.

The car had success at home too, in the Scottish, Welsh and RAC rallies. The SS Car Club ran its own meetings for owners at Donington, at which Bill Heynes and William Lyons would themselves take part. And one owner, Paul Marx, shipped his car to the East Coast of America, where he took part in road race events organised by ARCA, the Automobile Racing Club of America. Against an odd assortment of customised vehicles, Marx's Jaguar (reputedly the only one raced in the U.S. before the war) introduced America to the British car's potential, taking fourth place in an eight-mile hill climb up Mount Washington in New Hampshire.

In the SS Jaguar 100 everything came together completely for the first time – Lyons' design sense and business acumen, coupled with the engineering skill of Heynes and Weslake. But the growing success of SS Cars brought its own problems. Traditional ash-framed coachbuilding was simply too slow and painstaking a process to satisfy the demand which the new models were generating.

The company was still not commercially strong enough to commission entire prefabricated bodies, and the decision was taken to continue to use small, pressed-steel panels bought in from outside contractors alongside some made on site at the company's Foleshill works. But from now on they would be assembled onto a steel framework, which would itself be assembled from a number of sections. The scheme for the new body was designed by Lyons himself.

Sourcing so many components coming from so many suppliers was a risky strategy, and nearly proved to be the small company's undoing. When the first batches of elements arrived for the new all-steel body in late 1937, they didn't match up. The bodies simply couldn't be assembled, and production was held up for many months while the mess was sorted out through negotiation and a certain amount of legal pressure

Left: This SS Jaguar 100 dates from 1937. One of the most noted features on the car was the enormous Lucas headlamps.

Below: The 2½-litre engine of the SS Jaguar 100, with its new DHV cylinder head, when combined with the light weight of the car, enabled it to reach speeds of up to 95mph.

on the suppliers. Production of the all-steel Jaguars recommenced in the spring of 1938.

In the years following the launch of the SS100, the company continued to develop the car. The vehicle which Tommy and Elsie Wisdom had raced around Switzerland became a testbed for innovations by Bill Heynes. Heynes was ably assisted now by Walter Hassan, who had made a name for himself in a similar role at Bentley and at ERA (English Racing Automobiles).

Heynes and Hassan swapped the car's old 2.5-litre engine for the 3.5-litre version and positioned it further back on the chassis, while moving the axles further forward. The result for 'Old Number Eight' (as Jaguar enthusiasts know the car because of its chassis number 18008), was a speed of 125mph achieved at Brooklands in 1939. In all, Jaguar would build 118 of the 3.5-litre SS100s and with a speed of 60mph reached in under 11 seconds it was one of the fastest sports cars in the world. An Alfa Romeo 2500 might have been faster but it was £1000 more expensive than the £445 SS Jaguar 100.

There were other developments, including an astonishing one-off fixed-head SS100 coupé that was built for the 1938 London Motor Show. Stripped of its running boards and with a long tapering nose seemingly wedged between two huge front wings it bore more than a passing resemblance to the future XK120 fixed-head coupé.

But no sooner had William Lyons ironed out the problems of all-steel manufacture than another, more serious interruption to production came along. A few cars were produced in 1940 but soon the entire SS factory was turned over to the war effort. The Foleshill works were used for manufacturing and repairing aircraft frames, including Spitfire parts and Meteor jet fuselages, skills that would come in useful after the cessation of hostilities.

In March 1945, as victory in Europe drew nearer, Lyons addressed a meeting of shareholders of SS Cars, whose company name was now a problem: "Unlike SS, the name Jaguar is distinctive and cannot be connected or confused with any similar foreign name." The shareholders agreed in consequence to adopt a new company identity, Jaguar Cars Limited.

The car that made the company's reputation had now given the company its name. The pre-war range of Jaguar saloons was put back into production from September 1945, but in a new age of austerity the 100 was dropped.

Below: Only one SS Jaguar 100 fixed-head coupé was ever built, for the 1938 London Motor Show. In 1938 the company also introduced a 3½-litre engine for the sports car, propelling it to 104mph. Only 118 of this ultimate of pre-war sportscars were ever built.

JAGUAR MARK V

At the end of the Second World War, Britain was desperately short of everything. Every aspect of British industry, including the Jaguar works in Coventry, had been redeployed in the service of the war effort. Steel was in particularly short supply and the British government prioritised its allocation to manufacturers with strong export potential.

In order to secure a quota of the scarce material therefore, Jaguar belatedly began to turn its attention to the overseas market, which it had not given particular focus before the war. This failure was in part the legacy of Richard Taylor, an English dealer who had acquired the rights to market SS Cars in North America in 1935, but done very little to establish the brand there.

In search of a new and more proactive East Coast distributor, managing director William Lyons appointed Max Hoffman, an Austrian émigré, on the advice of his old friend Bertie Henly. It was Henly's large order for the Austin Seven Swallow back in 1927 which had encouraged Lyons to focus on cars rather than sidecars. Charles H. Hornburg Jr. was the man given responsibility for the West Coast. At the same time Lyons took the decision to sell off the sidecar business to one of his wartime clients, Helliwell Ltd. Helliwell, like the Jaguar works, had been involved in aircraft maintenance during the war. The new owner continued to produce Swallow sidecars until the late 1950s.

The greatest early challenge faced by Jaguar was the revival of the Coventry production line, which had been mothballed for five years. Recommissioning the machinery was costly and complicated, but by September 1945 the company was back in production, producing the saloon version of its pre-war SS Jaguar range in all three engine sizes (minus the 100).

As Lyons reshaped the company for its post-war future, he made an important acquisition; the tooling necessary to produce the 2.5- and 3.5-litre six-cylinder Standard engines now being used across the company's range of pre-war models. This gave him a valuable measure of independence, although Standard would continue to produce the 1.5-litre four-cylinder engine which Jaguar used. In late 1947 the saloon was joined by the drophead coupé, available with either of the six-cylinder versions. By then the first Jaguar models with left-hand drive were being shipped to America. Jaguar

JAGUAR MARK V

ENGINE inline six-cylinder

CAPACITY 3485cc

BORE X STROKE 82 x 110mm

COMPRESSION RATIO 7.2:1

POWER 125bhp

VALVE GEAR overhead camshaft

FUEL SYSTEM twin SU carburettors

TRANSMISSION four-speed manual

FRONT SUSPENSION independent wishbones, semi-elliptic springs, torsion bar, anti-roll bar

REAR SUSPENSION live axle, semi-elliptic leaf springs

BRAKES Girling, rod operated

WHEELS steel, bolt-on

WEIGHT 3640 lbs (1651 kg)

MAXIMUM SPEED 90mph (145kmh)

PRODUCTION 1005, 1948-1950

Left: The 1948 Jaguar Mark V was available as a drophead coupé (from 1949) as well as a conventional saloon, but it would go on sale for just three years.

was embracing the world stage at last.

Lyons was looking to the future. Compared to their American counterparts from Detroit the pre-war models were looking distinctly dated, and the Standard engines had been developed more or less to their limits. He set about designing a new, more powerful engine for the Jaguar range, the XK engine, intended initially for a new range of luxury saloons aimed at the export market. But although in 1948 the new chassis and engine were ready, the body wasn't – Pressed Steel, the supplier of the larger panels required for the new range (the future Mark VII), simply had too many orders and couldn't guarantee to fulfil Jaguar's until 1950 at the earliest.

A new saloon was needed to fill the gap, and at the London Motor Show in October 1948 Lyons presented the new Mark V. Admittedly the new model contained the old engine, and the new engine was contained in the shell of a concept sports car (the XK120). But the dual unveiling gave the Jaguar stand a bigger impact and turned the frustrating delay to its best advantage.

Nowadays the Mark V is rather unfairly dismissed as an interim measure, merely bridging the gap between the post-war SS Jaguars and the delayed luxury saloon.

It was overlooked in favour of the sensational concept car which sat beside it on the Motor Show stand. But although the Mark V used the old models' existing engines and gearbox it offered several interesting innovations. It used the robust new chassis design destined for the Mark VII. It was supported on smaller wheels, 16-inch down from 18-inch, fitted for the first time with hydraulic brakes and much improved suspension – for the first time, independent at the front.

Below: The saloon-bodied version of the Mark V. It was named the Mark V as it was supposedly the fifth of five prototypes.

Left: The front wheel hubcap emblazoned with the Jaguar logo that is still in use today. The 'Jaguar' badge on the bootlid would last through to the 1960s.

Below: The Mark V came with either the 2½-litre or 3½-litre 'Jaguar' straight-six, pushrod engine dating from the 1930s.

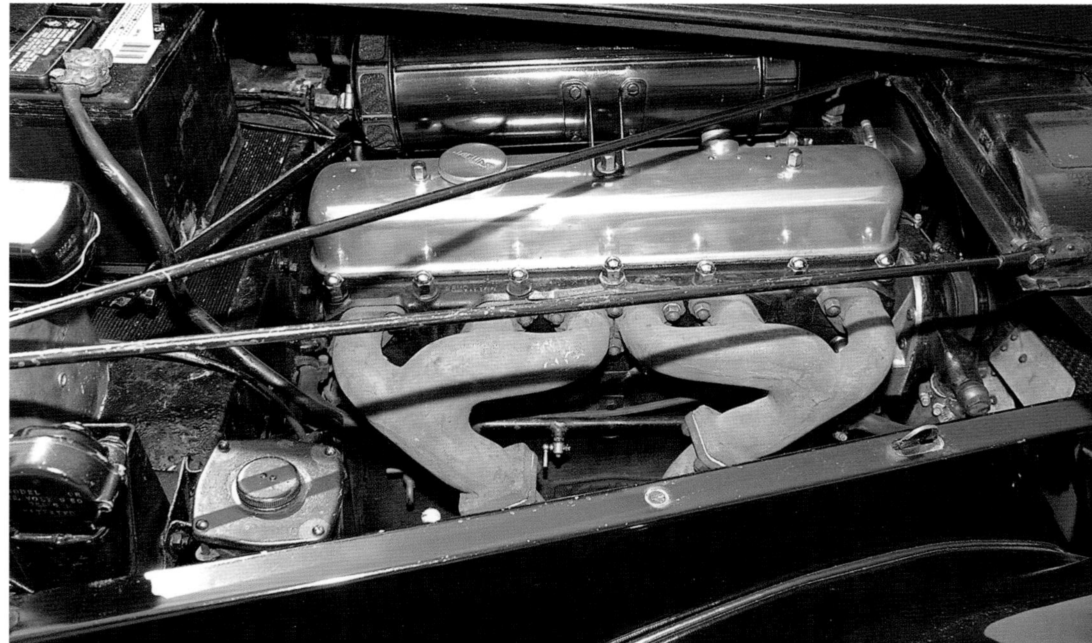

The body was clearly derived from its predecessor the SS Jaguar saloon and was still constructed from small plates on a steel frame. But in detail it was larger (and heavier) all round and almost every panel was new in form. The doors were wider, there was a softer, sleeker curve at the back of the roofline and down across the boot. It was the first of a discernible Jaguar saloon-car line.

With prominent chrome brightwork on the side windows, push-button door handles which sat in line with the waistline trim, headlamps faired into the bonnet and double bumpers front and back, the whole vehicle exuded luxurious elegance at a time when the post-war nation was much in need of both values. It had the aura of a Bentley but not the price tag.

Naming it the Mark V, associated it with the luxury car maker Bentley, whose own Mark V had appeared just before the war. Indeed pre-war Jaguars bore a more than passing resemblance in some details to their Bentley counterparts.

In 1949 Jaguar released a beautiful two-door drophead coupé version of the Mark V. Like the saloon it was available in both six-cylinder engine sizes, but aspects of its construction, not least the drophead itself, meant that it was produced in much smaller quantities than the hardtop. Its scarcity now and its success at the time in motorsports have made it much sought after by collectors.

If the Mark V had not been introduced as a stopgap model, one suspects that it would have enjoyed a much longer production life, based on its commercial and sporting success. As it was, its successor was finally ready to roll off the production lines in 1950, and the manufacture of the Mark V ceased in 1951.

JAGUAR XK120

The XK120 came about through an act of brilliant expediency rather than a planned strategy. The car was designed from scratch in a hurry as a showcase for another car's engine. William Lyons decided to produce a new sports car for the 1948 London Motor Show to help promote the new saloon, the Mark V and the planned Mark VII's radical new engine, the XK.

The breakthrough in engine power which Harry Weslake had delivered before the war brought home to Lyons the importance of the Jaguar machinery as well as its appearance. Good looks had to go hand in hand with good performance and the old Standard engines had been engineered to the limits of their potential.

Lyons and his engineers Bill Heynes and Wally Hassan set about creating an entirely new engine for the company, now assisted by Claude Baily who had joined Jaguar from Morris in 1940. It was the brainchild of Bill Heynes, who chose to make the engineering leap from tested, reliable, overhead pushrod valves to an overhead cam arrangement more commonly seen in pre-war racing cars. Hitherto such an engine had not been considered reliable enough or quiet enough for a road car, but Heynes pulled it off. In time it would become the workhorse, or rather the racehorse, of the Jaguar range.

It was complemented by a new Jaguar chassis, a tremendously rigid affair with new suspension offering Jaguar's most comfortable ride yet. Lyons intended to place chassis and engine at the heart of a new range of luxury saloons and coupés, the future Mark VIIs.

Delays in producing the bodywork for the Mark VII, and with the Mark V engineered to take the old 2.5- and 3.5-litre Standard engine, he had no showcase for Heynes' breakthrough straight six. The obvious setting for the powerful and innovative new engine was a sporty modern vehicle, something that reflected

Below and Right: Sir William Lyons' 'afterthought' for the 1948 London Motor Show was such a success that the American agent, Charles Hornburg, wanted to buy the entire first year's production run.

RJH 400

its potential speed and new approach. In a remarkable burst of creativity, Lyons designed and produced a sports car of outstanding beauty in only six weeks, and named it the XK120, based on the assumption it would reach 120mph.

This claim was an act of faith on Jaguar's part – the XK was completely untested. Lyons planned to use the concept car as a testbed for the engine, and hoped to sell a couple of hundred XK120s if it turned enough heads.

It was a two-seater body put on top of a shortened version of the new saloon chassis (No. 670001) and fitted with the 3442cc XK engine. There wasn't a straight line in sight, from the curved radiator grille at the tip of the long nose to the gently tapering boot. The rear wheel spats, which Lyons had introduced in the Mark V, were present here too. The cutaway line of the doors emphasised the red leather trim of the two-seater interior. This was no family car – the only thing rising above the bonnet was the windscreen. This was a car for adults to enjoy life at speed.

The XK120 was an immediate success. Charles Hornburg Jr. visiting from the U.S. wanted to buy the first year's entire production. While the London Motor Show was still running it became clear that the company would not be able to keep up with demand.

Right: At a time when many sports cars were quite basic, the cockpit of the XK120 showed a degree of luxury.

Top: The open boot of the XK120 complete with an original set of tools provided with each car.

Right: Although the new XK engine was revolutionary, and would become a Jaguar mainstay for 38 years, access to the installed unit was difficult in the XK120.

The original bodywork of the XK120 called for an ash frame with aluminium panels, which had to be built virtually by hand; but now William Lyons placed an order with Pressed Steel (the company already engaged to produce the body for the Mark VII) to make an all-steel body for the car, which could be more quickly assembled on the production line.

Until 1950, when Pressed Steel delivered, Jaguar worked flat out producing the wood-framed original, of which around 240 were manufactured. It quickly became a popular sports car in the USA. Hollywood star Clark Gable, was an early adopter and owned three in all, including the first ash-frame production car, chassis number 670003.

The overwhelming success of the XK120 persuaded Lyons to abandon plans for a cheaper version, the XK100. It would have been powered by a 2.0-litre, four-cylinder reworking of the XK engine, the XJ, a prototype of which stood alongside the original XK120 at the London show. But the XJ engine would

JAGUAR XK120

ENGINE six-cylinder

CAPACITY 3442cc

BORE X STROKE 83 x 106mm

COMPRESSION RATIO 8:1

POWER 160bhp

VALVE GEAR dual overhead camshafts

FUEL SYSTEM dual SU carburettors

TRANSMISSION 4-speed manual

FRONT SUSPENSION independent wishbone, torsion bar

REAR SUSPENSION live axle, leaf springs

BRAKES 4-wheel drum

WHEELS steel, bolt-on

WEIGHT 2912 lbs (1321 kg)

MAXIMUM SPEED 124.6mph (200.5kmh)

PRODUCTION 12,055, 1948-1954

have cost nearly as much as the XK to manufacture; and there seemed little point in bringing out a slower version of a road car whose main attraction was its speed.

In 1948, 120mph was an astonishing speed for a production car and there were those who doubted the truth of the claim. In May 1949 journalists were invited to an empty stretch of straight motorway at Jabbeke in Belgium, where a production XK120 managed several runs at over 130mph. In 1950 the steel-bodied version finally became available, with a few small changes to the lines of the wings and an extra 50kgs of weight from the heavier material, but with a performance much the same as the original.

In 1951 Jaguar added a fixed-head XK120 coupé to the family, at first with left-hand drive only, aimed firmly at the export market. The right-hand drive version for British roads did not appear until 1953.

If purists were afraid that the perfect aerodynamic lines of the original convertible would be spoiled by adding a hardtop, they needn't have worried. Because the cockpit shape was actually more aerodynamic than the open-topped, low-flying roadster, the fixed-head coupé could still reach comparable speeds.

The XK120's combination of performance and interior build won it the hearts and minds of drivers on the roads of America and Europe. Heynes' suspension solutions provided a most comfortable ride, and the power of the XK engine made for effortless driving. But there remained few public roads on which you could really open the throttle to the XK's full potential. For that, you needed a race track and it's fair to say that the XK120 took the motor racing world by storm.

Three months after the Belgian demonstration mile, three XK120s were entered in a one-hour production car race at Silverstone, where they took first and second place, and would have taken third too had one driver not suffered a puncture. William Lyons understood the value of racing success in generating production car sales, and for the 1950 season he lent six aluminium-bodied cars to six drivers who he thought had the potential to show the XK120 at its racing best.

Two went to Leslie Johnson and Peter Walker, who had already proved their mettle in taking those top

Below: Wire wheels became an option for the XK120 in 1951 when the fixed-head coupé was introduced. This right-hand drive version would only be available from 1953.

Top: The first 240 XK120s were built traditionally with aluminium bodies, but such was the demand that the move to a 'production' steel body was inevitable.

Above: The drophead coupé joined the XK120 range in April 1953. It came with a veneered dashboard.

two places at Silverstone. Another went to Tommy Wisdom, an old friend of Jaguar who had driven the SS Jaguar 100 to success in the Alpine Rally. Australian driver Nick Haines and Italian Clemente Biondetti each received one; and the sixth was lent to Ian Appleyard.

Appleyard had, like Wisdom, distinguished himself in an SS100, with a third place in the Alpine Rally of 1947. In the XK120 he became the first driver to complete the Alpine Rally without penalty in three successive years, 1950-1952. He was very much part of the Jaguar family – his co-driver on those Alpine runs was his wife, William Lyons' eldest daughter Pat.

The XK120 achieved high placings at many great races throughout the 1950s, often with Johnson or Appleyard at the wheel. Johnson drove the car in its first American win, at Palm Beach in January 1950. Four months later XK120s took first and second at Pebble Beach. The same year it was Wisdom's generosity in lending his car to a young driver at a meeting on the Dundrod track in Northern Ireland which gave that driver, Stirling Moss, his first major win.

Although the XK120 never won at Le Mans or in the Mille Miglia, it performed in those races with distinction, and encouraged Lyons to design a car that would bring Jaguar victory there. In the XK, Jaguar now knew that it had a big engine that could deliver results.

JAGUAR C-TYPE

The success of the six XK120s which Jaguar had lent to half a dozen racing drivers in 1950 only made William Lyons hungry for more success. In particular he wanted the laurels from the Le Mans 24-hour race, a test of endurance for man and machine since 1923, the oldest and most prestigious race in the motoring calendar.

Three XK120s had entered the race in 1950. Peter Whitehead and John Marshall finished in fifteenth place, Nick Haines and Peter Clark in twelfth. Leslie Johnson and Bert Hadley had been in second place and closing in on the leaders and eventual winners Louis Rosier and his son Jean-Louis when just three hours from the finish the Jaguar clutch failed and had to retire. To add insult to injury, two Aston Martin DB2 cars, Jaguar's great racing rivals, finished ahead of the two finishing XK120s.

Lyons knew he had a winner in the XK120. That much had already been demonstrated at other races. So the changes he imposed on it to transform it into the XK120C, or C-type, were relatively minor in mechanical terms – a new rear suspension arrangement with a single torsion bar replaced the production model's leaf springs, and a more precise rack-and-pinion steering system replaced the steering box. The XK powerhouse was given a tune-up, with raised compression, and some upgrades to the durability of

Left: A 1952 C-type in the colours of former independent Jaguar team, Ecurie Ecosse.

parts which would be tested to their limits over 24 hours – a new camshaft, a crankshaft damper, a better bottom end and high-spec bearings.

Most of the Jaguar team's efforts were directed towards improving the XK120's aerodynamics, under the direction of Malcolm Sayer, and in decreasing its weight. First to go were production frills such as carpets and door handles; the windscreen was made shallower and no longer interrupted by a centre post. Gone too were the rear wheel spats, for which there was no room over the more prominent hubs of the wire wheels.

A new lightweight tubular frame devised by Bob Knight supported an aluminium body made of as few pieces as possible. The bonnet and front wings were a single panel, hinging forwards for engine access. The characteristic radiator grille was also shortened, widened and accommodated nearer the ground. A single panel covered the rear end too, and the classic flowing lines of the sides were flattened – a sad loss to the visual delight of the car but an essential aerodynamic improvement.

Jaguar entered three XK120C cars (the C stood for Competition from which the name C-type would be derived) in the 1951 Le Mans. If 1950 had been a portent of things to come, 1951 was a triumphant if flawed demonstration of Jaguar's presence. The Moss/Fairman C-type led the Biondetti/Johnson C-type from early on, and Stirling Moss himself posted the fastest lap of the race. But problems with the Jaguar oil feed forced both cars to withdraw after less than 100 laps of the nearly nine-mile circuit.

All the team's hopes now rested on the experienced heads of Peter Walker and Peter Whitehead, whose C-type seemed unaffected by the failures of the others. And after 24 hours, the chequered flag came down with Peter Whitehead crossing the line after completing 267 laps, nine ahead of his nearest rival and – even more satisfyingly – at least two places ahead of the nearest Aston Martin DB2. Whitehead and Walker's victory was won over 2244 miles (3610km), 60 miles more than the second-placed Talbot Lago, with an average speed of 93.5mph. The quality and power of the XK engine had now been brought to the world's attention.

Determined to repeat Jaguar's success in 1952, Lyons again entered three C-types in the Le Mans race. For the first time they were identified as a Jaguar works team rather than simply Jaguar-backed individual driver entries. There were further 'droop snoot' modifications to the aerodynamics of the cars, intended to squeeze extra speed from them. The racing department feared the capabilities of the Mercedes-Benz 300SL.

Jaguar team manager W.E. 'Lofty' England assembled a strong roster of drivers, but the restyled body of the C-type was its undoing. It generated less downforce on the rear, so that at speed the tail actually lifted and made steering difficult. Worse, the

Above: Two views of Ian Stewart's C-type, chassis no. XKC 006. It was the first C-type raced by Ecurie Ecosse and co-driven by fellow Scot Ninian Sanderson, it won sports car races at Snetterton and Silverstone in 1953.

Above right: Chassis no. XKC 015 was built in 1952 and competed in races in the U.S.A. in 1953 in the hands of Masten Gregory.

JAGUAR C-TYPE

ENGINE six-cylinder

CAPACITY 3442cc

BORE X STROKE 83 x 106mm

COMPRESSION RATIO 9:1

POWER 204bhp

VALVE GEAR dual overhead camshafts

FUEL SYSTEM two SU carburettors

TRANSMISSION Moss four-speed manual

FRONT SUSPENSION independent via wishbones and torsion bars, telescopic dampers, anti-roll bar

REAR SUSPENSION single transversely mounted torsion bar

BRAKES automatically adjusting drums

WHEELS wire, bolt-on

WEIGHT 2240 lbs (1016 kg)

MAXIMUM SPEED 145mph (233kmh)

PRODUCTION 53, 1951-1953

slight alterations necessitated a new configuration of the cooling system, which reduced its efficiency. Overheating was the result, and two cars pulled up with blown gaskets. The third car, having quickly replaced the half-size C-type radiator with a full-size XK120 one, avoided the overheating problem but had to retire after mechanical problems led to a loss of oil pressure.

For the 1953 Le Mans race, Jaguar undid the aerodynamic damage of 1952, reverting to the 1951 styling. Instead, vital savings in weight were made by using thinner aluminium in the body, thinner steel in the frame and finer electrical wiring. The rigid metal fuel tank was replaced with a rubber bag. Disc brakes were fitted to the cars for the first time, which gave them a crucial advantage in being able to brake later than those relying on drums. And the twin SU carburettors were replaced with a trio of Weber ones, raising the XK engine's power to 220bhp.

'Lofty' England brought back the same drivers who had been let down by their cars in 1952. Disaster almost struck during practice sessions, when Rolt and Hamilton were disqualified for accidentally driving with the same number and at the same time as another Jaguar – Belgian team Ecurie Francorchamps were also entering a C-type. Rolt and Hamilton were already in the bar drowning their sorrows when England told them that he had managed to get them reinstated. As the race got underway the crew tried to ply Hamilton with much-needed coffee, which (he claimed in his autobiography) he rejected in favour of brandy. Thus anaesthetised, he felt no pain when later in the race a bird flew into his face while he was travelling at 130mph, breaking his nose.

Moss and Walker took an early lead, only to fall back when a clogged fuel filter starved the engine. Two hours into the race, Rolt and Hamilton's car took over at the front and from then on it was a battle between Ferrari and Jaguar until, with a few hours to go, Ferrari faded and the Jaguars powered through.

At the finish it was Hamilton first, Moss second and Stewart fourth. The Ecurie Francorchamps C-type completed the race in ninth position. Rolt and Hamilton became the first men to drive Le Mans at over 100mph, maintaining an average speed of 106mph. They had driven an extraordinary 304 laps, 2540 miles (4087km) in 24 hours.

The 24 Hours of Le Mans 1953 are in many ways Jaguar's finest. The complete domination of the race by the C-types of the Jaguar team made them the envy of the motor-racing world. The following year the team rolled out the C-type's successor; but Ecurie Francorchamps ran theirs again in 1954 and came fourth. There was life in the older model yet.

6
JAGUAR

JAGUAR D-TYPE

The D-type was developed with expertise acquired during the C-type's successful assaults on Le Mans. Its objective was the same – victory at Le Mans – and Jaguar set about its goal with a confidence backed by two C-type wins in three years.

The XK engine remained a reliable powerhouse and relatively small changes were made there to better its performance. Dry-sump lubrication was introduced to reduce the engine's height off the ground, which gave the whole vehicle a lower centre of gravity. A new cooling system replaced the hasty solutions required to overcome the C-type's early failings in that department. The power output was boosted to 250bhp with a better manifold.

Once again Jaguar called on aerodynamicist Malcolm Sayer to squeeze more speed out of the already slippery body. Sayer had joined the company from the Bristol Aeroplane Company and brought his deep understanding of aerodynamics to bear, employing a wind tunnel to test out his designs. The vehicle's central monocoque construction was his innovation – a weight-saving technique used in aeroplane design in which the alloy body got its strength from the form of the panels instead of relying on an underlying framework.

With the D-type, Sayer was allowed to abandon altogether the classic, tall Jaguar dropnose radiator which the C-type had still employed. Instead he aimed for as low a profile as possible, starting with a shallow but wide grille. To lower it further he tilted the engine slightly to the left, by an angle of 8.5 degrees. The single-piece bonnet and forward wings he lowered even further – so the engine had to be accommodated in an asymmetric bump within the front panel.

Sayer's design helped push the D-type's speed up so successfully in trials, that he was forced to make one last addition. It became the D-type's defining characteristic for the general public: a tailfin, running back from behind the driver's head, to overcome instability at high speed. The new car had the lean, hungry, predatory look of a shark.

The D-type made its debut at Le Mans in 1954.

Below: Mike Hawthorn with his characteristic Herbert Johnson helmet with visor, which he wore rain or shine. Hawthorn had just brought the D-type home for a grim victory in the 1955 Le Mans race. The car still bears its Le Mans number today, and pertinently, extra mirrors.

6
1000
MIGLIA

9
774RW

JAGUAR D-TYPE

ENGINE six-cylinder

CAPACITY 3442cc

BORE X STROKE 83 x 106mm

COMPRESSION RATIO 9:1

POWER 250bhp

VALVE GEAR dual overhead camshafts

FUEL SYSTEM triple twin-choke Weber carburettors

TRANSMISSION four-speed synchromesh

FRONT SUSPENSION independent via wishbones and torsion bars, telescopic dampers, anti-roll bar

REAR SUSPENSION single transversely mounted torsion bar

BRAKES disc

WHEELS peg-drive alloy

WEIGHT 2460 lbs (1116 kg)

MAXIMUM SPEED 175 mph (282 kmh)

Work finalising the fin-enhanced bodywork of the car had been so last-minute that the three cars' paintwork was only completed after their arrival trackside. Under Lofty England's team management, Jaguar fielded three experienced pairs of drivers: Tony Rolt and Duncan Hamilton, and Stirling Moss and Peter Walker, who had taken first and second place in C-types the year before; and Jaguar veteran Peter Whitehead, teamed this year with relative newcomer Ken Wharton.

Jaguar's closest rivals that year were Ferrari. The British team had the edge in stability, but the Italians had a 4.9-litre engine and unbeatable acceleration and braking. It was billed as a contest between brain and brawn; the wind-tunnel-developed sophistication of Britain versus the raw power of Italy.

During the race it was Jaguar who clocked the faster times, thanks to their aerodynamic superiority. But all three D-types were plagued with blocked fuel filters in the early stages; and the Moss/Walker vehicle, which at one stage was lying in third place, was forced to retire with brake trouble. Overnight, Whitehead/Wharton also pulled out with gearbox failure. Ferrari had faced similar problems and now both teams were reduced to just one car.

Hamilton and Rolt went all-out in defence of their title, eating away at the Italian marque's lead until their D-type trailed the 375 Plus by only 90 seconds. But in the dying moments on a drying track the Ferrari, driven by Argentina's José Froilán González, pulled ahead. Ferrari won it with a margin of less than one lap.

Brute force might have won over beautiful bodywork, but the D-type had demonstrated its potential, clocking up over 170mph on the Mulsanne straight, where Ferrari had only managed 160. Only a month later Whitehead and Wharton drove their D-type to victory in the Rheims 12-hour event.

Back in Coventry, the car's aerodynamics were further improved with a new long-nose profile. The engine and suspension were placed in a new, separate frame; and to compensate, the car's weight was reduced by replacing the bolt-on rear bodywork with a stronger, lighter, stressed panel. Engineers worked minor miracles with small modifications to the engine including larger valves, wringing a powerful 270bhp from the far-from-production XK.

Jaguar again entered three cars at Le Mans in 1955. Lofty England assembled two new teams: Mike Hawthorn with Ivor Bueb, and Don Beauman with Norman Dewis. Dewis joined Jaguar in 1952, not to race but as the company's chief test driver. In 1953 he drove a modified XK120 at 172mph on the Jabbeke straight in Belgium where the model had first proved its mettle in 1949. If anyone at Jaguar knew how to handle one of the company's cars, it was Dewis.

The race started well, with Jaguar, Ferrari and Mercedes neck and neck in the early stages. The Ferraris faded with mechanical problems, leaving Hawthorn in the lead. Hawthorn, lapping and passing Lance Macklin's Austin Healey, then had to move in front of Macklin's path when he noticed that he had been called in to the Jaguar pits, which were then alongside the main straight. Macklin swerved left to avoid Hawthorn, and moved into the path of Frenchman Pierre Levegh driving a Mercedes 300SL. Levegh himself swerved left to avoid Macklin but clipped the Austin Healey's rear and was launched into the air.

Above: Power output from the D-type's engine was gradually increased over the years with fuel injection. Shown here is the normally aspirated engine of a 1955 car.

The Mercedes cartwheeled over an embankment into the watching crowd and hit a concrete stairwell, where it disintegrated. The suspension, engine and other debris flew on through the mass of spectators while the fuel tank burst into flames. Levegh and 83 members of the public died. One hundred and twenty were injured. Macklin himself was unharmed. It remains to this day the worst ever accident in motor racing history.

As emergency services attended to the dead and dying, the race was allowed to continue, to avoid the prospect of departing spectators obstructing ambulance crews. By midnight it was Mercedes, Jaguar, Mercedes at the front. But Mercedes-Benz's board of directors met urgently during the evening. They decided to withdraw their team from the event as a mark of respect, and at 2am the 300SLs were retired. Jaguar, invited to do the same, declined.

With the Mercedes-Benz cars out of the race, Hawthorn and Bueb had no competition, and drove to a hollow victory 14 hours later, to the scorn of the French press. Rolt/Hamilton retired with gearbox failure late in the race and Beauman/Dewis did not finish following a separate accident. A D-type of Ecurie Francorchamps finished in third place.

Although Hawthorn was widely blamed for the disaster, an enquiry found that the primary cause was an outdated and inadequate track layout approaching the pits straight. In future the pit signalling would be moved to the exit of the Mulsanne corner. The main straight had not been designed for the speeds now being achieved by participants, and was substantially widened in time for the 1956 race.

Hawthorn's and England's decisions during the race may well have been affected by an earlier tragedy. John Lyons, only son of Jaguar's founder and managing director William Lyons, was killed in a road accident on his way to the 1955 Le Mans race. He was 25 years old. John had recently started at Jaguar as an apprentice, and his father hoped that he would learn about the company from the shop floor up, before taking over the reins as son and heir. For William Lyons personally, and for the close-knit Jaguar family, the death of John must have thrown a dark shadow over the forthcoming race, and affected the judgement of Hawthorn, England and others.

Mercedes withdrew altogether from motorsport at the end of the season. The works Jaguar team returned to Le Mans with D-types in 1956, although only the car of Hawthorn and Bueb finished the race, in sixth position. First place however was taken by a D-type entered by David Murray's Edinburgh-based Ecurie Ecosse and driven by Ron Flockhart and Ninian Sanderson. It seemed, unsurprisingly, that the fight had for the moment gone out of Lyons and Jaguar, and the company followed Mercedes in withdrawing from motorsport at the end 1956.

The company's interest in the D-type may have faded as it stopped racing and focused on new production vehicles, but other teams continued to believe in its power and potential. Even without direct Jaguar backing, the 1957 Le Mans saw the D-type's best ever showing: five were entered, and five finished, all in the top six places. Ecurie Ecosse D-types took the top two spots, and Ron Flockhart again drove the winning car, this time co-driven by 1955 winner Ivor Bueb.

JAGUAR XK140

For a model originally intended as a short-run testbed for an engine, the XK120 was the progenitor of a long and distinguished line of Jaguar cars. Its immediate successor, the XK140, lacked the shock of the new, but it was a better car, offering a more mature ride.

Launched as a 1955 car at the 1954 London Motor Show it couldn't fail to have a more orchestrated arrival than its predecessor. It was immediately available in three different versions – a two-seater roadster (described officially by Jaguar as an 'open two-seater'), as well as a fixed-head coupé and a drophead coupé,

Below: The XK140 was launched in 1954 with the XK120 bodyshape but a host of different refinements.

the latter two models with two very small rear seats for occasional use on short journeys. The car had gained three inches more legroom, was two and a half inches wider, and under the fixed-head roof there was one and a half inches more headroom. It still had the styling of the XK120, but in an altogether more luxurious, more comfortable body.

To gain more room inside, the reliable XK engine was moved forward three inches and the bonnet also shortened. The dashboard was raised an inch and, in the coupé versions, pushed forward a further two inches. The boot of the coupé was also truncated to make room for the rear seating, and the battery (now just one 12v unit instead of two 6v ones) was moved from behind the seats to under the bonnet.

Outside there were other changes that were more or less cosmetic. The radiator grille was now one piece, with fewer, wider bars – seven, down from fifteen. Behind it a new fan drew in extra air to combat overheating problems which the XK120 had experienced in stationary traffic. The bumpers were fatter too, matching those which the company had already introduced in the Mark VII saloon. There were newly designed rear lights and the front wings now sported indicators. A chrome strip ran down the centre of the bonnet and across the boot, where it included a proud (if slightly misleading) new badge, 'Winner – Le Mans, 1951-3'.

In the early 1950s Britain was still emerging from the austerity of the immediate post-war years, and Jaguar's eyes were firmly fixed on the more affluent American marketplace. Many of the improvements were therefore aimed firmly westwards, not least the thicker bumpers, designed to protect it from parking-

Left: Much of the exterior trim was altered on the new XK140, notably the bumpers and radiator grille. The 190bhp engine that had been an option on the XK120 now became standard and was repositioned to improve handling.

Right: Jaguar's advertising images for the XK140 range. All three variants of the model were immediately available. Note the standard models all had pressed-steel, bolt-on wheels.

THE 3½ LITRE JAGUAR XK140 OPEN 2-SEATER

THE 3½ LITRE JAGUAR XK140 DROPHEAD COUPE

THE 3½ LITRE JAGUAR XK140 2-3 SEATER COUPE

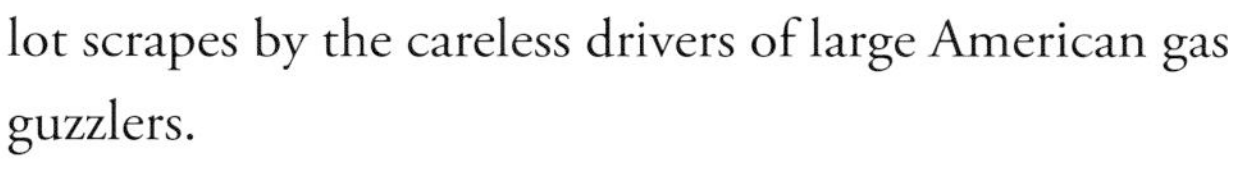

lot scrapes by the careless drivers of large American gas guzzlers.

Optional extras included a Laycock de Normanville overdrive switched from the dashboard, which delivered the same sort of benefit as a fifth gear. Better fuel economy and smoother cruising were of greater interest on the long, open American highways than the winding country lanes of England. In the last year of the production run, XK140s also offered the option of a fully automatic Borg-Warner transmission, the first Jaguar sports model to do so.

The standard pressed-steel wheels came with the rear spats which Jaguar had first presented on the XK120. Most of the XK140 production was assembled with left-hand drive and shipped to America – in total only 73 right-hand drive versions were built – and most of the American exports chose instead the spatless option of wire wheels. The torsion bars and antiroll system introduced in higher-spec XK120s were now standard in the XK140, and suspension was now by telescopic shock absorbers instead of the old lever-arm set-up.

The emphasis was on comfortable driving. The roadster, or 'open two-seater' had a primitive hood with plastic side windows, but the drophead had wind-up glass. The roadster's trim was leather and leatherette, including the dash; but the drophead had all-leather upholstery with walnut veneer on the doors and dashboard. The roadster's windscreen was detachable but the drophead screen had a fixed surround in the body colour, with quarterlights.

In terms of performance the XK140 did not match its predecessor, let alone achieve the implied 140mph top speed. The larger body put paid to that, and in tests it never managed more than 125mph. The relocation of the engine put more of the weight on the front wheels – in fact distributing it almost equally between the two axles – which improved its handling on straights but not its stability on curves. The C-type version of the XK engine was offered as an option to increase the XK140's power, which added 20bhp to the output and required twin exhausts. In this form the car became the XK140SE (for Special Equipment), known as the XK140MC in North America.

It was undeniably a powerful road car, but a road car nevertheless, and rarely entered in competition. It only occasionally made racing appearances: a roadster from early in the production run, chassis number 5 and registration TNK 140 sold by Jaguar's oldest supporter Henly's of London, was soon upgraded with the C-Type engine and driven at Goodwood, the only XK140 roadster ever to appear there. William Lyons' son-in-law Ian Appleyard, for many years an enthusiastic racer of XK120s, switched to a fixed-head XK140, registration VUB 140, and took part in the RAC Rally in 1956.

It may not have had a glittering racing career, but on the road the XK140 was unbeatable. Its production car rivals, the Mercedes-Benz 300SL and BMW's 507 had the pace but could not compete on price; the Aston Martin DB 2/4, the XK140's exact contemporary, had neither the power nor the price advantage. In 1955 a new XK140 would cost around £1600.

Left: The rear seats of the coupé made it a 2+2, but Jaguar described it as a '2-3 seater'. The drophead coupé also had rear seats but they were dispensed with entirely in the roadster.

JAGUAR XK140

ENGINE six-cylinder

CAPACITY 3442cc

BORE X STROKE 83 x 106mm

COMPRESSION RATIO 8:1

POWER 190bhp

VALVE GEAR dual overhead camshafts

FUEL SYSTEM twin SU carburettors

TRANSMISSION four-speed manual (automatic optional)

FRONT SUSPENSION telescopic, independent wishbone, torsion bar

REAR SUSPENSION telescopic, live axle, leaf springs

BRAKES Lockheed drums

WHEELS pressed steel, bolt-on

WEIGHT 3136 lbs (1422 kg)

MAXIMUM SPEED 121mph (195kmh)

PRODUCTION 2790, 1954-1957

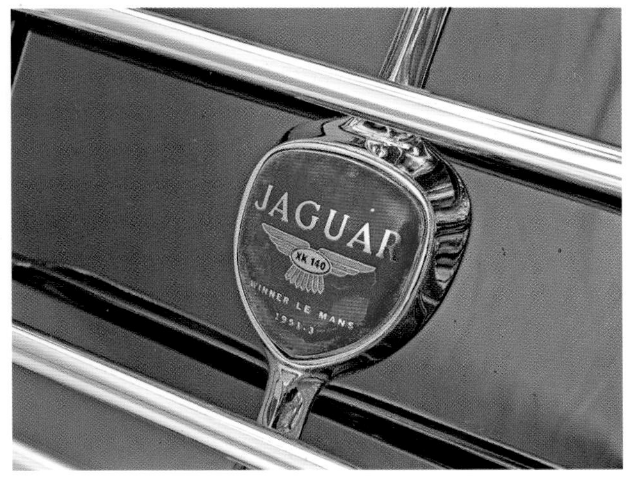

Left: When it was launched, the model badge on the boot lid could only boast two wins at Le Mans.

JAGUAR MARK VII

After the stopgap Mark V gave the public something new from Jaguar after the war, Pressed Steel were finally able to deliver the body panels for the Mark VII in early 1950. This was the luxury model which William Lyons hoped would conquer North America for Jaguar, and he gave it a separate launch in New York only a few months after its London debut. In fact the delay to its production caused by Pressed Steel's full order book was to the car's advantage in Britain. Post-war restrictions were at last easing and petrol rationing, which had begun in 1939, ended in the year of the Mark VII's launch. Life's little luxuries were becoming available once again.

Not that the Mark VII was in any sense a little luxury. Although its fuel consumption was not bad for its time, around 17mpg, it was a huge beast, weighing in at nearly two tons (1800kgs) and measuring over 16 feet in length. It was, in terms of handling, a car less suited to the small lanes and tight corners of European town and country than the big roads of North America.

Wherever it appeared, however, it turned heads. Where the Mark V had been little more than an adaptation of the dated pre-war SS Jaguars, the Mark VII was an all-new, post-war visualisation of elegance. It is said that Lyons worked closely with his model makers and draftsmen at the design stage, using his

instinctive eye for a graceful curve to direct the car's form.

Its beauty was not only skin deep. The interior finish was luxurious, and its mechanical performance was one of controlled power and comfort.

The Mark V had been limited by the old Standard Motor Company 2.5- or 3.5-litre engine. The Mark VII became the first Jaguar saloon to run with the magnificent XK engine, which had made its debut so triumphantly with the XK120 roadster.

Once installed in its rightful place,the XK engine did not disappoint. Virtually unchanged it powered Jaguar cars of all kinds from saloon to racer from its XK120

Below: The XK-powered Mark VII saloon was Jaguar's first 100mph saloon. It truly delivered the Jaguar slogan, 'Grace...Space...Pace'.

Left and Below: The Jaguar Mark VIIM was introduced after four years, as a facelift model. Fog lamps were moved outside the body shell, their place taken by a pair of horn grilles.

unveiling in 1948 to its final production use in the XJ6 S3 of 1987. Although a four-cylinder version was conceived, only the six-cylinder version was produced beyond prototype. Designed for endurance, it was housed in a solid iron block, and the steel crankshaft turned on seven bearings. In keeping with Jaguar's philosophy of beauty and strength together, the twin-overhead-camshaft cylinder head was of polished aluminium.

The Mark VII's interior was grandly appointed. More than five hides of leather were used in the upholstery, and over 50 sections of walnut veneer decorated the doors and dashboard. The engine sat five inches further forward than in the Mark V, giving ample legroom for five adult travellers. The boot had a larger capacity for their luggage than any other car on the road at the time.

If the Mark VII embodied Jaguar's taste for power and style, its name also reflected William Lyons' shrewd sense of marketing. With the Mark V he had aligned

JAGUAR MARK VII

ENGINE six-cylinder

CAPACITY 3442cc

BORE X STROKE 83 x 106mm

COMPRESSION RATIO 8:1

POWER 160bhp

VALVE GEAR dual overhead camshafts

FUEL SYSTEM twin SU carburettors

TRANSMISSION 4-speed manual (automatic optional)

FRONT SUSPENSION independent wishbone, torsion bar

REAR SUSPENSION live axle, leaf springs

BRAKES servo drums

WHEELS steel, bolt-on

WEIGHT 3864 lbs (1753 kg)

MAXIMUM SPEED 103mph (166kmh)

PRODUCTION 20,938, 1950-1954

Left: The interior was an ocean of leather and walnut veneer, giving it the feel of the more expensive Bentley.

himself with the luxury market occupied by marques such as Bentley: the declaration of war had sabotaged the launch of Bentley's own Mark V, of which only 17 were manufactured. The Bentley Mark VI arrived in 1946, and with the Jaguar Mark VII (there was no Jaguar Mark VI) Lyons deftly suggested his new car's superiority to the Bentley.

With its launch, it embodied attributes which Jaguar would later adopt as its marketing slogan: 'Grace, Space and Pace'. It was welcomed with open arms by the luxury car market, and in the first six months of its life, the Mark VII brought in $30 million from export sales to the United States alone.

If the sheer size of the Mark VII made it awkward on narrow European roads but at home on American highways, its size and power lent themselves surprisingly effectively to motorsports. A succession of Mark VIIs completely dominated the Daily Express Production Saloon event at Silverstone, winning it five years in a row from 1952 to 1956.

In 1954 Jaguar rolled out an upgraded version of the Mark VII, the Mark VIIM. It was another

Left: Concealed in a door panel, the Mark VIIM came supplied with its own grease gun.

Above: The Mark VIIM engine was uprated from 160bhp to 190bhp and gearbox ratios made closer.

Right: The suspension was modified to reduce the amount of roll in cornering.

stopgap revision, while the company worked on the forthcoming Mark VIII. It featured mostly cosmetic improvements to the bumpers and wheel trim. The old trafficators had been replaced at the front by wing-mounted indicator lamps, and wing mirrors were now standard. More significantly Jaguar's engineers managed to squeeze another 30bhp out of the XK engine thanks to high-lift camshafts and a close-ratio gearbox.

In an original Mark VII Ian and Pat Appleyard had come within one second of victory in the 1953 Monte Carlo Rally. In 1956, with a Mark VIIM, the all-Irish team of Ronald Adams, Frank Bigger and Derek Johnstone finished the job, taking first place ahead of a Mercedes-Benz 220.

Jaguar might have done the double in the Monte Carlo Rally too, but in 1957 the rally was cancelled, a victim of the petrol shortages resulting from the Suez Crisis. A combined Israeli-French-British operation to take control of the Suez Canal backfired and resulted in, among other things, the re-introduction of petrol rationing in Britain in December 1956. With serious shortages, fuel coupons were not available for such luxuries as motorsports. That sense of scarcity also affected the public's appetite for gas-guzzlers like the Mark VII, whose sales now slowed. Smaller, more efficient cars were available and although rationing was lifted again in May 1957, sales never recovered. Around 10,060 Mark VIIMs were produced, half as many as the original Mark VII, but more than either of its successors the Mark VIII and Mark IX.

Such had been the early sales success of the Mark VII that Jaguar had to relocate to larger premises to keep up with the demand. Luckily such premises were available. During the war the British government had built a series of 'shadow factories', which drew on the expertise of neighbouring peacetime industries such as car manufacture to assist in the production of military machines and weaponry. In 1951 Jaguar moved into the shadow factory in Browns Lane in Coventry, which had been used during the war for aircraft sub-assembly. Browns Lane would remain the company's headquarters for the next 54 years.

Below: An original Mark VII, distinguishable by the fog lamps which remain in the bodyshell. The steering wheel is non-standard.

JAGUAR XK 150

The XK150 was the last and most radical reworking of the XK sports car series. The styling throughout implied width rather than length – stability rather than speed. The radiator was short and squat compared to earlier Jaguars, although the grille returned to the thinner struts of the XK120. The bonnet was broader, opening right up to the wings, its slimness gone in favour of easy access to the engine compartment.

Gone too was the racy curve of the wings, which had been a prominent feature of Jaguars since the arrival of the XK120, an echo of the shape of the running boards going back to the SS100 and the SS2. Instead, they continued in no-nonsense straight lines across the doors from front to rear wheel arch, before rising slightly to meet the boot.

The doors were thinner, the seating inside correspondingly wider to accommodate a broader range of clientele. The dashboard and doors were now trimmed not in sleek walnut veneer but in leather – padded leather in the new top-roll above the instruments. The seats themselves, although

Below: The rear view of the XK150 shows the style convergence between the fixed-head sports car and the new Jaguar Mark 1 saloon.

upholstered in pleated leather to Jaguar's usual high standards, still didn't offer much in the way of side support, especially on corners at speed. Like the XK140, the coupé versions of the new car included rudimentary rear seats. These were little more than cushions below and behind, in the space occupied in the XK120 by a storage locker. Jaguar was right to describe them as 'occasional'.

As if to confirm that its racing days were behind it, the XK150 originally appeared in only fixed-head and drophead coupé versions. The roadster followed only a full year after the launch, and even that model had been softened for comfort. The roadster's hood, hitherto a basic affair offering little more than crude protection from the elements, was much improved. The plastic side windows were now replaced by wind-up glass ones; and the whole enclosure was much easier to put up, and stowed very neatly behind the seats –

Left: While the XK140 had continued the lines of the XK120, the XK150 launched in 1957 was noticeably different from its predecessor with a higher wing line and a curved one-piece windscreen.

unlike the canopy of the drophead edition, which sat rather untidily above the boot when not in use.

All models came with a single-piece curved windscreen, an innovation in the XK range although it had been introduced a year earlier in the Mark VIII saloon. The XK150 fixed head had a much larger rear quarterlight than the XK140, and a broader rear window too. So in the glasswork too, width was emphasised.

As with all Jaguars however, the look was only half the story. Some things had not changed. The choice of transmissions still included a Moss four-speed manual gearbox, with or without overdrive, or Borg-Warner three-speed automatic. Under that wide bonnet the XK 3.4-litre engine was the standard issue, still delivering 190bhp.

Top left: The radiator grille of the XK150 reverted to the old XK120 style.

Above left: The boot-lid badge was now able to boast five Le Mans wins, yet again confirming the importance of the French race to Jaguar.

But a Special Equipment version, the XK150SE, was given a new B-type head which combined the exhaust valves of the C-type with the XK's smaller-diameter inlets. The B-type head pushed the car's output up to 210bhp. Disc brakes (developed jointly by Jaguar and Dunlop) were also part of the SE package, so the XK150 was no longer prone to the brake fade of earlier drum-fitted models. Now you could drive the car to its limits secure in the knowledge that you could rely on its ability to pull up when required.

Still, the emphasis on width and comfort left the XK150 frustratingly slower than its predecessors: the standard XK150 had a top speed of only 123mph. William Lyons set engineer Harry Weslake to work on the problem. Weslake had worked wonders with the Standard Motor Company engines which Jaguar used before the war, and now he did it again. By straightening the supply tubes he ensured a better mixture for the engine at higher revs. Pushing the compression ratio up to 9:1 and adding a third SU carburettor transformed the performance of the machine, now delivering 136 mph from 250bhp.

Weslake's solution was a response to the challenge of the V8 engines now being developed in America. The resulting vehicle, the XK150S, appeared in 1958, initially available only as a roadster and aimed very much at the American market. Fixed and drophead models carried the S engine from 1959.

In the competitive market of the high-performance car, even Jaguar could not rest on the laurels of its recent run of Le Mans successes; and the company was already working on a longer-term solution, one it hoped would keep it ahead of the field in the same

Below: The interior of the XK150 was modernized with veneer disappearing from the dashboard.

way that the mighty XK engine had. As the XK range of vehicles was nearing the end of its life, the new 3.8-litre engine was being developed for a different car altogether, the E-type. But in the short term they tried it out in late production XK150s.

The results, launched in 1959, were impressive, especially in the XK150 3.8S, which also included Weslake's tweaks for the XK150S. Boasting 265bhp it could go from 0 to 100 in under 20 seconds, and reached nearly 140mph. Only around 270 of the 3.8S version were produced before production of the XK150 came to an end, but Jaguar must have been very encouraged by its performance.

Taking into account all the different engine and body combinations, by the end of its time the XK150 had been made available in 12 different models. Despite the attractions of the 3.4S, 3.8 and 3.8S, it was the original roadster, drophead and fixed-head coupé that generated the greatest sales. Sales of the basic fixed-head model accounted for a third of all 12 versions, underlining the fact that this was a car for comfortable road-driving rather than exhilarating speed.

The birth of the E-type spelled the end for the XK150 and the whole XK programme. The startlingly modern E-type form made the old car look very old. The XK150 quickly fell from favour. Its value plummeted and surviving vehicles fell victim to neglect and corrosion. Many were driven until they fell apart. It wasn't until the mid-1970s that its reputation began to recover. The E-type, which had ousted the XK in people's affections, was then approaching the end of its own life, to be superseded by the XJ-S series. As an appreciation of classic cars began to emerge, the XK150 was a prime candidate for rehabilitation, and its rarity in good condition only helped to make it more desirable.

JAGUAR XK 150

ENGINE in-line six-cylinder

CAPACITY 3781cc

BORE X STROKE 87 x 106mm

COMPRESSION RATIO 8:1

POWER 265bhp

VALVE GEAR dual overhead camshafts

FUEL SYSTEM twin or triple SU carburettors

TRANSMISSION four-speed manual or three-speed automatic

FRONT SUSPENSION independent wishbones, torsion bars, anti-roll bar

REAR SUSPENSION live axle, semi-elliptic leaf springs

BRAKES 4-wheel disc

WHEELS wire or pressed steel, bolt-on

WEIGHT 3136 lbs (1422 kg)

MAXIMUM SPEED 136mph (219kmh)

PRODUCTION 9382, 1957-1961

JAGUAR XKSS

Although the D-type had been designed solely for the purpose of winning at Le Mans, a production version was launched in 1955, for the small market of private enthusiasts who could afford the hefty £3878 price tag. Le Mans required that cars entered in it should, at least in theory, be the prototype for a production car with a run of at least 100 vehicles. In the end 68 were produced over the next two years; but despite the car's Le Mans successes it proved a hard sell for the open road. By the end of 1956, although some dealers were offering discounts of as much as £1000, 25 production D-types had not been sold.

Four of them were broken up for parts, but Lyons saw in the remaining 21 an opportunity to enter a previously closed U.S. market. The racing D-type had not been eligible as a road vehicle under the regulations of the SCCA (Sports Car Club of America). With some judicious adaptation, the unsold machines might be

converted into acceptable roadsters for the American sports car market.

One of the SCCA's requirements was (like Le Mans) a production run of at least 50 cars, and Jaguar duly announced the 'new' car, the XKSS, in January 1957. SS stood for Super Sports. The D-type's distinctive dorsal fin was abandoned, along with the dividing bar between driver and passenger. Otherwise, the changes were mostly additions, and mostly cosmetic ones in the name of driver comfort. A basic hood was added, a full-size windscreen, a passenger door and side windows too. In the racing D-type there was no room for storage. Even the production version of the D-type had no glove compartment, and what should have been the boot was occupied by the spare wheel; so a rear luggage rack was a standard accessory on the XKSS.

Brightwork was incorporated in the new look, with a chrome windscreen frame and chrome outlines to emphasise the D-type's aerodynamically faired headlights. New chrome corner bumpers were a necessary feature for road-going use and these were cut-downs from the double bumpers of the Mark VIII and Mark I saloons already in production. This bumper styling worked so well that it was later carried forward, slightly amended, onto the E-type.

The XK engine retained many of the racing improvements made to it, including the triple Weber carburettors, and it still delivered 250bhp. The whole vehicle had all the power of the D-type and carried only a little more weight and wind resistance…and luggage. It has been described as the first supercar, and with its pedigree of Jaguar racing vigour and roadster

Left: Only 16 of the intended 50 Jaguar XKSS cars were ever built, before a fire at the Browns Lane factory destroyed nine cars and the tooling required to complete the rest.

Bottom: The engine was a full 250bhp D-type providing 150mph performance. It really was a road-going Le Mans car.

style it should have been a winner, especially in North America.

Only a month after the launch of the XKSS, a disastrous fire in February tore through the Jaguar factory, destroying much work in progress and much of the tooling required for the XKSS. Other models were also affected, including the Mark I and the new XK150. Only the hands-on support of the Jaguar workforce got the production lines back up and running at all. The damage could have been much worse, but the XKSS was not revived. Only 16 of the projected 50 of the new model had been produced.

Only one of the 16 remained in Britain. One went to Hong Kong where in time it won the Macau Grand Prix. Two were exported to Canada, where one at least enjoyed a long and successful career in hillclimbs. The remaining dozen all went to the United States, and two of those ended up racing in Cuba before Fidel Castro came to power. The Cuban cars were brought back to Britain in the 1980s, but in 2010 a reunion of 12 of the original 16 took place at Pebble Beach as part of Jaguar's 75th anniversary celebrations.

Nowadays fashion designer Ralph Lauren is among the lucky few to own an XKSS. His car was present at the Pebble Beach Concours d'Elegance. Historically the most famous previous owner of an XKSS was film actor, and star of the movie *Le Mans*, Steve McQueen, who bought his in 1958. McQueen, the car's third owner, hired hot rod customizer Tony Nancy to replace the original red leather upholstery with black, and changed its factory white paintwork to British Racing Green. McQueen sold the car eleven years later, but such is the bond between a man and his Jaguar that in 1977 he bought it back for double the price he got in

1969. Nicknamed the Green Rat, McQueen's machine is now part of the Petersen Motor Museum collection in Los Angeles.

One XKSS has been through more than its fair share of changes. The very first XKSS had started life as a production D-type before being converted and shipped to the US as a demonstration model in the New York showroom. After a chequered career in racing, it was restored in the 1970s to its original D-type form by a Japanese owner before later being re-converted back (or perhaps forwards) to an XKSS specification.

Fast forward to 2016 and following on from Jaguar's successful construction of the remaining six chassis numbers assigned to the E-type Lightweights in the early 1960s – 18 were assigned, but only 12 were built – Jaguar announced a similar proposition for the XKSS. Nine new XKSS cars would be hand-built by Jaguar Classic to the exact specification as they appeared in 1957, replacing the cars lost in the Browns Lane factory fire. As with the Lightweight E-types, the bespoke cars carried a price tag in excess of £1 million for delivery in 2017.

JAGUAR XKSS

ENGINE six-cylinder

CAPACITY 3442cc

BORE X STROKE 83 x 106mm

COMPRESSION RATIO 9:1

POWER 250bhp

VALVE GEAR dual overhead camshafts

FUEL SYSTEM triple twin-choke Weber carburettors

TRANSMISSION four-speed synchromesh

FRONT SUSPENSION independent via wishbones and torsion bars, telescopic dampers, anti-roll bar

REAR SUSPENSION single transversely mounted torsion bar

BRAKES 4-wheel Dunlop disc

WHEELS pressed steel, bolt-on

WEIGHT 1800 lbs (816 kg)

MAXIMUM SPEED 149mph (240kmh)

PRODUCTION 16 in 1957 and 9 in 2017

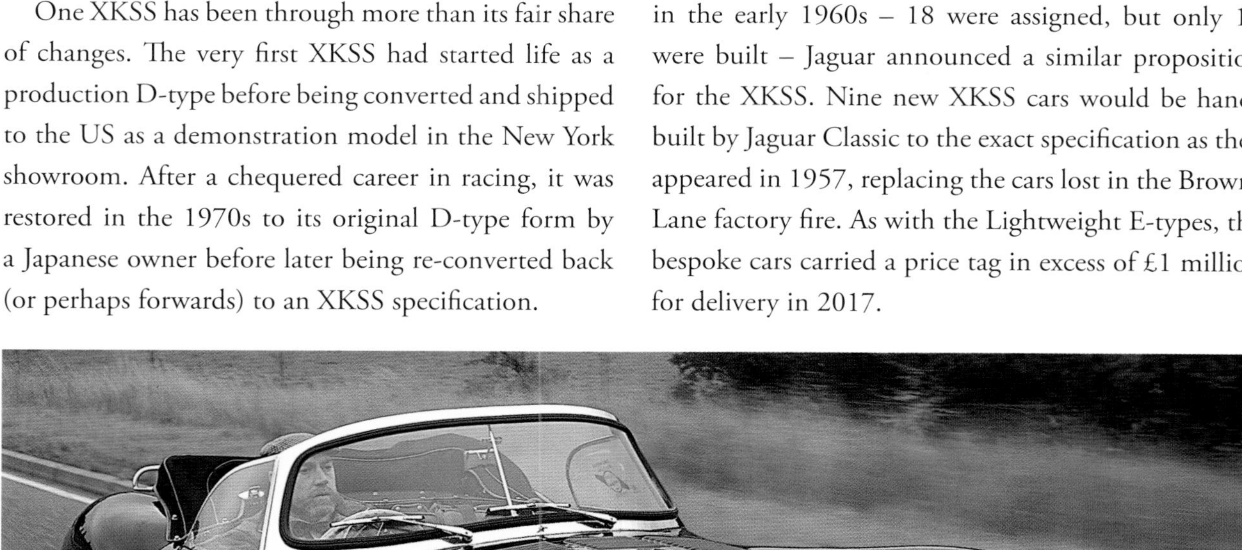

Left: Steve McQueen and wife Neile Adams with 'the Green Rat' and their car collection circa 1967. He is reputed to have racked up 80,000 miles in an XKSS.

JAGUAR MARK 2

Jaguar had announced a new compact saloon in September 1955. It was highly significant because it would be the company's first volume car. The body was of unitary construction and with no separate chassis to insulate the bodyshell from road noise, much attention was paid to refining the suspension. Bob Knight, under the guidance of Bill Heynes, had engineered a front sub-frame mounted on the body via rubber bushes to carry the new coil-spring suspension, while the cantilevered rear leaf springs were flexibly mounted within steel boxes.

Power was provided by a new short-stroke version of the XK engine, still six cylinders, but only 2483cc. With its modern bodyshell in a unique 'teardrop' shape styled by Lyons exuding great refinement, and all the familiar Jaguar leather and walnut veneer inside, the 2.4 saloon was one of the most sophisticated saloons yet produced. It sold in the U.S. for $3700, a Jaguar for substantially less than the Mark VIIM.

As usual, criticism from the States centred on power, or lack of it, and it was remedied by the option of the Special Equipment 210bhp 3.4-litre engine. Equipped with this engine unit the 3.4 saloon supplanted the Bentley Continental at the world's fastest four-seater

production car. What would become known as the Mark 1 made way for the all-conquering Mark 2.

It was only with the arrival of the Mark 2 saloon in 1959 that these preceding models became known as Mark 1s. As the humble Jaguar 2.4 and 3.4, the Mark 1s had supplied the market for a four-door family saloon for four years. The model was extremely successful, and more than 35,000 were manufactured during its relatively short life.

The Mark 2 was launched as the age of the British motorway dawned. The use of an Arabic numeral 2 rather than the Roman numerals of earlier Marks was a deliberate indication that this was a modern car, a car of progress. The Mark 2's mission was to rectify a number of problems which the Mark 1 displayed, and to ensure that Jaguar held its ground in the competitive medium saloon market well into the 1960s.

Like the Mark 1, the Mark 2 benefited from unitary construction. This monocoque body style, introduced by Malcolm Sayer for the racing D-type, used body panels built in such a way that they supported themselves instead of requiring a separate frame. The Mark 1 had been the first road Jaguar to adopt this technique, and the plant to make it in Jaguar's new Browns Lane premises required an investment of over a million pounds, a huge sum in 1954. So it was no surprise that, to avoid unnecessary further investment, William Lyons (now Sir William, having been knighted in 1957) was determined to use as much of the existing tooling as possible for the Mark 1's successor.

By dispensing with a frame, there were great advantages in weight and strength. Lyons found other

Below: The Mark 1 (in black), with noticeably thicker window frames, was originally known as the Jaguar 2.4 litre saloon and subsequently the 3.4 litre saloon. With the arrival of the Mark 2 (bottom), these became known as Mark 1 cars.

Right: Designed for economy rather than performance, the 2.4-litre straight-six XK engine propelled the Mark 2 to 60mph in a sluggish 17.5 seconds.

Left: For the Mark 2, the large speedo and rev counter were relocated from the centre console, as in the Mark 1, to directly in front of the driver. Jaguar's traditional four-spoke steering wheel was also replaced with a more modern design that featured a single cross-spoke.

ways to cut back on framework too. The full-frame doors of the Mark 1 were heavy affairs; now they became half-frames, with lightweight chrome window frames bolted on.

The new car retained the design lines of the Mark 1. The rear lights were given new housings, and at the front, the sidelights were moved onto the top of the wings to match their position on other Jaguars. Indicators now occupied the lower positions. Between them, the radiator grille now sported a broad, chromed central rib, which nowadays acts as a quick and easy way of telling Marks 1 and 2 apart.

One very obvious difference was the greatly increased expanse of glass in the Mark 2. The old model had attracted criticism for being claustrophobic when fully occupied: the new one had 18% more window area than before, including over nine inches more on the sides of the compartment. This was achieved by the thinner upper frames on the doors; by stretching the characteristic D-shaped rear quarterlights; and by using narrower pillars at the corners of the front and rear screens. The rear screen itself was now almost a wraparound, seven inches wider than the Mark 1's and three inches deeper in its curve. Not only was there more light and space for the occupants; there was more visibility all round for the driver.

Inside there were further improvements. The front seats had deeper upholstery, with picnic tables built into their backs for the benefit of the backseat passengers. The floor was covered in Wilton carpet, and once again the dashboard and door caps were of highly polished walnut veneer. The dash itself had a new arrangement, moving the major dials (the speedometer and the rev counter) directly in front of the driver at last. To

Above: A John Coombs conversion of a 1962 3.8-litre Mark 2 with bonnet straps, multi-slatted louvres and protective grilles over the headlights.

MHT 704F

maximise their visibility the four spokes of the old steering wheel were reduced to a single cross-spoke.

The old Jaguars had notoriously bad internal heating, and the Mark 2 attempted to address that problem with a new system which included hot air ducts to the rear. It must be said that this new solution was only partially successful.

The new saloon came, as the old one had, with a choice of 2.4-litre and 3.4-litre versions of Jaguar's ever reliable XK engine. But now they were joined by a third option, a 3.8-litre edition, previously available only in the larger Mark IX saloon. This made the Mark 2 the fastest production saloon in the world, with powerful acceleration up to 125mph. All three engines had the option of automatic gears and, from the end of 1960, some much-needed power steering – without that help, the Mark 2 was heavy at slow speeds and tended to understeer.

Although Jaguar had withdrawn from team racing three years earlier, many private drivers and teams took full advantage of the 3.8's performance. One of the most important improvements in the Mark 2 was to fix a handling problem which had plagued the Mark 1 and which may have had tragic consequences for veteran Jaguar racer Mike Hawthorn.

The 3.4 version of the earlier Mark had also achieved success in motorsport, notably with old hands Hawthorn or Stirling Moss at the wheel. Sadly, in January 1959, it was in a customized Mark 1 that Mike Hawthorn died in an accident on a public road in Surrey. The precise cause of the accident was never established. It seems likely that Hawthorn was road-racing his friend and fellow Jaguar veteran Peter Walker who was in a Mercedes 300SL; but the crash happened on a bend, and one of the known weaknesses of the production Mark 1 was its rather flighty cornering, the effect of a narrow rear track. The Mark 2 was given a wider rear track, concealed behind new, lower rear skirting. The higher roll centre of the Mark 2 also increased its stability.

The 3.8's racing success made it notoriously popular as a getaway car for thieves and villains, and consequently with the police forces trying to catch them. But the Mark 2's most famous police driver is a

JAGUAR MARK 2

ENGINE in-line six-cylinder

CAPACITY 3781cc

BORE X STROKE 87 x 106mm

COMPRESSION RATIO 8:1

POWER 220bhp

VALVE GEAR dual overhead camshafts

FUEL SYSTEM twin SU carburettors

TRANSMISSION four-speed manual (overdrive and automatic options)

FRONT SUSPENSION independent semi-trailing, double wishbones, coil springs, telescopic dampers, anti-roll bar

REAR SUSPENSION live axle, semi-elliptic springs with twin radius arms, telescopic dampers

BRAKES 4-wheel disc Wheels pressed steel bolt-on (wire option)

WEIGHT 3360 lbs (1448-1524 kg)

MAXIMUM SPEED 125mph (201kmh)

PRODUCTION 83,980, 1959-1969

Left: Despite a change of badge at the rear, the 240 still maintained the radiator grille badge of the original Mark 2.

Above: At the end of its production run the Mark 2 was rebadged the 240 for the 2.4-litre engine and 340 for the 3.4-litre engine model.

fictional one, TV's Inspector Morse. Just as Inspector Maigret is synonymous with the Citroen Traction Avant, so the Mark 2 is associated with Morse. In Colin Dexter's original novels, Morse drives a Lancia; but actor John Thaw, who played Morse from 1987 to 2000, insisted that his character drive a British car, a red 2.4-litre Mark 2.

In need of more factory space, and having had planning permission for a new factory turned down, Jaguar bought the Daimler Motor Company in 1960. Daimler were a British manufacturer that had purchased the rights to use Gottlieb Daimler's authoritative name on motor cars from the turn of the century and were also based in Coventry. But it was a company lacking the clearsighted vision and entrepreneurial drive of Sir William Lyons.

The last car launched by Daimler as an independent marque was an uncharacteristic sports car, the SP250, and so Jaguar inherited what by common consent was the ugliest new model at the 1959 New York Motor Show. It was a poorly designed, badly made roadster, a straw clutched at by a company with more experience of building tanks and buses than of motorsport. But it had the great advantage of being powered by a V8 2.5-litre engine designed by Daimler's engineering wizard Edward Turner.

Although Sir William bought Daimler principally for its factory sites, Jaguar needed to launch a new Daimler to rescue its automotive reputation. In 1962 Lyons' pragmatic, cost-saving solution was to market the Daimler V8: effectively a Mark 2 body with the 2.5-litre version of Turner's engine. Inside, the 'new' car exhibited an even higher standard of luxury than the Mark 2, although outside the only immediately discernible difference was the radiator grille, whose head was now furrowed instead of smooth.

Daimler's reputation as the limousine maker for kings and presidents allowed the V8 to occupy the more mature, sedate end of the Mark 2's medium saloon market – with its plush interior and the quieter, smoother ride of Turner's engineering it appealed to doctors, lawyers and other professionals who might have been discouraged by the Mark 2's racier image.

During the 1960s the Mark 2 was upgraded in many details to keep it competitive in an increasingly fierce market. As sales tailed off, the 3.8 ceased production in 1967, and as the company prepared to launch its new saloon the XJ6, the remaining two Jaguar models were rebranded as the Jaguar 240 and 340. The Daimler V8 became the Daimler 250 and all three cars suffered some unfortunate economies of interior trim to keep them keenly priced. A switch to plastic upholstery, for example, made the Mark 2 the first Jaguar in which leather did not come as standard.

All the versions of the Mark 2 had a long production life, and were Jaguar's (and Daimler's) most successful cars to date. Between them over 100,000 were made and sold, and the Daimler V8 remains that badge's most popular model of all time. Sir William Lyons still had the magic touch.

JAGUAR E2A

As a new decade dawned in January 1960, there seemed little prospect that Jaguar were about to return to the scene of their great sporting triumphs. They had won the Le Mans 24 Hours race five times in the 1950s; twice with the C-type and three times with the tail-finned D-type. However after the D-type of Ecurie Ecosse had beaten the works team and won the race in 1956, Jaguar withdrew from sports car racing, leaving it to the privateers.

William Lyons, it seemed, was now focused on turning sporting success in the word's greatest motor race into commercial success and the D-type had only been seriously aimed at the smooth tarmac of the Sarthe as opposed to the more bumpy tracks it would encounter in the rest of the Sports Car Championship.

After the tragedy of 1955, when Pierre Levegh's Mercedes had been launched into the crowd, the Le Mans organisers, Automobile Club de l'Ouest (ACO), had applied a 2½-litre engine limit to prototype cars for 1956. However, Jaguar's existing D-type was regarded by the French club as a standard 'production car', as Jaguar had plans to build 100 similar cars, the qualification number for a 'production car'. The company knew it was going to be difficult finding 100 customers willing to buy a car with the specification and expense of a D-type and in the end resolved to transform the excess D-types into the XKSS.

Jaguar Competitions superintendent Phil Weaver wrote to Chief Engineer Bill Heynes in December 1955: "I feel we would not want to put through another 100 cars, with such limited use as the present D-type… in which case it might be advisable to consider making the prototype car in such a manner that it would have at least limited demand in a better equipped version, such as the Porsche, for purposes other than racing." Thus the future route for development of the E-type had been mapped out five years before its eventual emergence. It would be a competition car capable of rapid evolution into a road car.

The first attempt at the D-type's replacement was given the project name E1A and designed by Sayer. It was similar to its racing counterpart but had fully independent rear suspension instead of the live-axle set-up. Mounted under the bonnet was a straight-six, 2.4-litre XK-Series engine.

The second version of this car, dubbed E2A, was begun on January 1st 1960, finished by February 27th, and on the 29th was already being tried out at the Motor

Right: The E2A in the Briggs-Cunningham colours leads a Triumph TRS into the braking zone for the Mulsanne corner at the 1960 Le Mans. Even in 1960 spectators were allowed to sit or walk by the side of the track.

Left: A studio shot of E2A as it was catalogued for sale by Bonhams, now with its rear fin removed.

Industry Research Association (MIRA) test track at Lindley. In appearance it looked like a smoothed-out D-type, with a typical E-type front end and a D-type rear end. The bodywork had exposed riveted panel work, which gave the car an aeronautical look. If it were to race at Le Mans that year it had to comply to the new regulations which required windscreens to be a minimum of 25cm (9.8 inches) high with an operable luggage compartment. The idea by the governing body had been to promote the conversion of sports prototypes into road-going vehicles.

The car had been fitted with an experimental five-speed gearbox, but Jaguar test driver Norman Dewis helped convince engineers that it should be replaced by the more reliable D-type four-speed box. With that installed he reached 161mph on the main MIRA timing straight. After just seven days of testing E2A (the 'A' was for the aluminium-skinned monocoque) was then taken, still in bare unpainted aluminium, trade-plated 'VKV 752', to the Le Mans test weekend on Saturday, April 9th 1960.

For the race it was painted in the Cunningham team racing colours of white with two parallel blue stripes. American multi-millionaire Briggs Cunningham had hired two American drivers, Dan Gurney (Eagle racing founder and inventor of the ubiquitous Gurney flap) and veteran multiple SCCA Champion Walt Hansgen to share driving duties. In practice Gurney had been struggling with the wayward handling of the car. It had been set up with the rear wheels 'toe out' i.e. both wheels pointing slightly outward from straight-on.

Below: Outside the factory at Browns Lane, the car looks from another age set against other utilitarian post-war vehicles.

When Gurney finally convinced the engineers to re-align the wheels so they were 'toe in', with both rear wheels pointing slightly inward, the wayward handling of the car was transformed.

It was an inauspicious start to the race. Gurney, had already collided with a Ferrari 250GT in practice warranting hasty repair attempts to the nose, and as early as Lap 3 Walt Hansgen was back in the pits with an injection pipe split, a familiar problem for the team at Le Mans. As the race unfolded, E2A proved it was quick down the straight, clocking the fastest speed on the Mulsanne despite its heavy chassis. By 8pm it had hauled itself up to tenth place, but it was now only running on five cylinders. At 1.40am a burned piston and head gasket failure proved too much for the engine and it was retired.

After Le Mans, E2A was returned to the Jaguar factory, where it was fitted with a 3.8-litre engine. The intention was that the Cunningham team could run the car in American SCCA races to promote the forthcoming launch of the E-type, or as it was known in the USA, the Jaguar XKE. It was raced on both seaboards with newly crowned Formula One World Champion Jack Brabham taking the wheel at the Los Angeles Times Grand Prix at Riverside, while Bruce McLaren also raced the car.

Its short racing career complete, it was shipped back to Coventry and unlike most racing thoroughbreds was adapted to test the Dunlop Maxaret anti-lock braking system, or WSP – 'Wheel Slide Protector'. The E2A still retains a dashboard button marked 'SHOT FIRING – PULL' to chalk-mark the road during testing. It was then put into storage until 1966. After that it was painted in British Racing Green, the tail fin removed and sent round the MIRA test track as a decoy while the factory tested its great new Le Mans hope, the V12 XJ13.

At this point, the factory could see no further use for it and the prototype was due to be cut up and scrapped. Veteran racing photographer Guy Griffiths heard about its impending doom. Together with his daughter Penny he had assembled a collection of important Jaguars, and most significantly, Penny's future husband Roger Woodley worked for Jaguar, maintaining customers' competition cars and heard of E2A's likely demise.

Roger persuaded former racing manager 'Lofty' England, who had become CEO of Jaguar Cars by that time, that E2A should be saved for the Collection. It was sold with the proviso that it could not be raced and sent off to the Jaguar body shop to have the bumps and knocks straightened out along with a respray back to the white and blue of its Cunningham colours. Sold without an engine, the owners eventually managed to obtain a 3.0-litre aluminium block as originally used at Le Mans.

After 40 years in the family, E2A was sold at auction by Bonhams at Quail Lodge in 2008 for $4,957,000, the highest sale ever recorded for a Jaguar at auction. Not a surprising sum for the original E-type Jaguar.

Below: The dashboard instruction 'SHOT FIRING - PULL' above the gearstick indicates that E2A was used as a factory test hack before it was earmarked for scrap.

Left: The stars of the Bonhams auction in 2008 were a D-type and E2A. Together they were expected to fetch $10 million.

Below: Jaguar design chief Ian Callum shakes hands with the proud new custodians of an important part of Jaguar history at the Pebble Beach Concours d'Elegance in 2011.

JAGUAR E-TYPE SERIES 1

Above: The E-type used many D-type design features such as a monocoque centre section, an engine carried by a projecting framework and a one-piece bonnet pivoting forward.

Very few cars are genuinely iconic – the Volkswagen Beetle, the Citroen DS, the Austin Mini; to which can be added the Jaguar E-type.

A direct descendent of the D-type, an immediate successor to the XK150, the E-type was initially produced only for export and presented for the first time at the Geneva Motor Show in March 1961 to instant acclaim. Even Enzo Ferrari was forced to admit that it was the most beautiful car ever made.

After Jaguar shut down its racing team in 1956, Sir William Lyons had instigated a search for a new roadster. The XK150 harked back to the 1948 design of the XK120, now nearly ten years old and starting to show its age. It was not a racing machine. Malcolm Sayer, Jaguar's wind-tunnel expert, was given a free hand to design as near perfect an aerodynamic form as possible, a shape fit for the future.

There were a couple of early prototypes; E1A which never left the factory and E2A, which raced at Le Mans in 1960 in the colours of the American team of Briggs Cunningham. Lyons felt that, although he was looking for a roadster, the rigours of racing would better test the new project's systems. The E2A ran with a 3.0-litre XK engine, later upgraded to a 3.8, and retained the D-type's trademark tailfin behind the driver's headrest. At the front, however, the wings kept a much lower profile and were part of the one-piece forward section of the body. It was clearly an E-type in waiting.

One of the first demonstration E-types almost kept Sir William Lyons waiting. At the 1961 Geneva Motor Show 150 of the world's motoring press were gathering at midday at a Geneva restaurant to get their first sight of Jaguar's new sports car – which wasn't there yet. Jaguar's PR man, Bob Berry, had asked test driver

Norman Dewis to bring the car out overnight from England.

There were already two E-types in Geneva for the launch; one was on the motor show stand plus a preproduction coupé, registration 9600HP, that was giving passenger rides around Lac Leman. Dewis had been testing an E-type convertible (77RW) at the MIRA test track in Warwickshire when he got the call to make the 700-mile trip. There were no British motorways on his route, although on the open European roads in the dead of night Dewis was able to put the E-type through its paces and reach speeds of over 140mph. Dense fog and heavy morning traffic on the approach to Geneva ate into his schedule.

He showed up at the restaurant with less than 20 minutes to spare. Lyons was reportedly furious with his staff for cutting it so fine, but the reception which the

Left: A Swiss-registered Series 1 E-type. For its Geneva launch an E-type convertible was driven overnight from England, reaching its destination with twenty minutes to spare.

car got soon overcame his anger, and in photographs of the event he is grinning from ear to ear. Of all the Jaguars, the E-type was his favourite.

At its launch, the new 3.8-litre E-type was available as either a fixed-head coupé or a roadster. Wire wheels and disc brakes were standard, and the dashboard reflected the new driver-friendly arrangement of the recently unveiled Mark 2. The steering wheel was wood-rimmed and three-spoke. Visibility was excellent, with a wraparound windscreen and, in the beautifully tapering fixed-head, exaggerated versions of the classic Jaguar D-shaped rear quarterlight. Both versions had wind-up side windows, and the open-top model had a simple fold-down hood. Two narrow bucket seats held driver and passenger tight on fast corners, and the trim was all of leather and carpet, with no veneer.

The fixed-head coupé had an early version of a hatchback's third door, hinged at the side to give access to the luggage space. The roadster, or in Jaguar parlance 'open top two-seater', had a considerably reduced boot capacity by the storage of the hood. It had the option of a chrome luggage rack. Long, tusk-like chrome bumpers protected and emphasised the curves of the

JAGUAR E-TYPE SERIES 1 & 2

ENGINE inline six-cylinder

CAPACITY 4235cc

BORE X STROKE 92 x 106mm

COMPRESSION RATIO 9.1

POWER 265bhp

VALVE GEAR dual overhead camshafts

FUEL SYSTEM triple HD8 SU carburettors

TRANSMISSION four-speed manual

FRONT SUSPENSION independent wishbones and torsion bars

REAR SUSPENSION independent lower wishbone, upper driveshaft link, radial arms, twin coil springs

BRAKES Dunlop discs with vacuum servo

WHEELS wire, bolt-on

WEIGHT 2744 lbs (1246 kg)

MAXIMUM SPEED 149mph (240kmh)

PRODUCTION 38,389, 1961-1968

Left: The fixed-head coupé had an early version of a hatchback's third door, hinged at the side to give access to the luggage space. This would be used again on the XK and the F-TYPE would also have a hatchback.

CKM 40

corners front and back. All exterior lights had chrome trim. The E-type's headlamps were set deep into the vestigial wings; their glass coverings preserved the aerodynamic integrity of the body, but tended to dim the beams, and to mist up in damp weather.

The body was of monocoque construction, manufactured by the Pressed Steel Company. It was so well designed, so rigid, that even without the upper body the open-top model needed no additional reinforcement. A fibreglass top was however available to transform the roadster into a saloon, with a shorter rearscreen drop onto the boot than the cabin of the coupé, and without the coupé's third door.

Under the cover, the front suspension was the same as the D-type's. But at the rear, some exciting new thinking had gone on, with a new system of independent suspension using coil springs and wishbones. In its 3.8-litre version, the 13-year-old XK block was still more than equal to the new car, which *The Motor* magazine tested to 149mph – admittedly with a works-tuned machine which would have outperformed any showroom model.

The roadster went on sale for £2098, the fixed-head coupé for sixty pounds more, around two-thirds of the price of the Aston Martin sports car, the DB4. An initial run of just 250 E-types was planned, but that plan was very soon abandoned: Jaguar could not keep up with the orders which poured in. There is a story told that in 1962, TV boss Lew Grade wanted Jaguar to lend his company an E-type for Roger Moore to drive in his new TV series *The Saint* – but Lyons turned him down on the grounds that they were selling them as fast as they could make them. The Saint would have to make do with a Volvo P1800 instead.

The very earliest production models featured an external latch for securing the bonnet, which could not be opened without the aid of a special tool. Replacing this with internally placed locks was one of the few changes made to the E-type in its first three years. Another was to dish the floorpan to give more legroom.

Then in 1964 came the first major technical changes. Jaguar improved the electrics with a new fan to cool the radiator and a new alternator-generator. At the same time the 3.8-litre engine was upgraded to 4.2 litres, in a longer block, with a new camshaft holding better bearings. The new workhorse gave the same power, speed and acceleration as the old one, but with increased torque.

A new all-synchromesh gearbox replaced the creaky old Moss box which was the E-type's original specification. And better brakes came via Lockheed in-line boosters, replacing the Kelsey Hayes servos of the early production, which were slow to respond at low speeds. All in all, the new 4.2 E-type was praised by motoring publications as the fastest, most-powerful

Below: Early versions of the E-type came with an aluminium dashboard and a laminated wood steering wheel.

Left: Jaguar estimated that demand would be 20 cars a week, but after its debut in Geneva and in New York a month later, orders were coming in at 120 a week.

Below left: The sealed headlamp buckets were finished in silver. In some restored models the bodyshell paint scheme has been applied.

production car they'd ever tested.

In 1967 the so-called 'Series 1½ E-type' was produced with a number of technical changes made to the original car in response to new U.S. regulations about emissions – different carburettors and recycled gases were the principle technical adaptations, and from outside the biggest difference was the removal of the headlamp coverings, a move which brightened the beam and was eventually incorporated in European models too. In the States, the E-type had taken on the branding initiated by the XK120, XK140 and XK150 and was known as the XK-E.

JAGUAR XJ13

By the early 1960s, the big, front-engined GT cars had reached the end of their days at Le Mans. Two lightweight E-types contested the 1964 race but they were significantly slower than the Daytona Cobra and the Aston Martin in the GT field, and neither finished the race. The XJ13 of 1966 was Jaguar's stillborn attempt to return to competitive racing with a fully-fledged works prototype car aimed to compete against the Ford GT40 and the Ferrari P4.

It was a project of chief engineer Bill Heynes who had overseen the introduction of the XK engine. It was to be powered by a mid-engined, 5-litre, four-cam V12. Jaguar had considered manufacturing an engine of this size as early as 1950. The initial plan had been to develop it as a racing engine, then adapt it for road car use; whereas the XK engine was built for road cars and subsequently adapted for racing. The engine design, commonly attributed to Claude Baily, was two XK six-cylinder engines on a common crankshaft with an aluminium cylinder block. The first engine ran on the test bench in July 1964 and would reach an output of 502bhp. Only seven of these experimental 5-litre V12s were made and only two to XJ13-specification with gear-driven camshafts.

The design of the car was from the pen of aerodynamicist Malcolm Sayer, who had created the C-type, D-type, and E-type. Jaguar test driver Norman Dewis, who was to have one of the biggest impacts on the shape of the XJ13, recalls that Sayer had a habit of drawing a full-size version of the cars he designed on the wall of his office and sometimes in chalk on the floor.

Sayer manufactured a scale model of the XJ13 and would take it to a wind tunnel test facility at Farnborough to check on aerodynamic flow around the vehicle and try to spot areas of turbulence and drag. When the full-size body had been assembled, he would attach small, three-inch strands of wool to the bodywork and follow test driver Norman Dewis around one of the MIRA (Motor Industry Research Association) test tracks, which was located close to the Jaguar factory, to check that the full-size prototype accorded with the model he had perfected.

At around 1248kg the car was heavy for a racer and despite Sayer's beautifully sculpted body, it needed the five litres of engine to make it fast. The project was developed on a part-time basis and by the time XJ13 was ready to race, its design had become obsolete against new cars from Ferrari, Ford and the Porsche 917. It also became the victim of a change in Le Mans engine regulations. Prototype cars were limited to engines of three litres. To run cars with larger engines, manufacturers had to homologate 50 examples as production cars (a figure later reduced to 25). Although there had been plans to build a number of XJ13s, as noted in an internal Jaguar memo which referred to 'the first car only', the company did not intend to make 50 and despite the car having been further developed by sports car drivers David Hobbs and Richard Attwood, the project was shelved.

During testing in 1966, Norman Dewis had reached speeds of 175mph in the XJ13 and managed to set a new lap-record time on the MIRA circuit. He would get to drive the car again five years later but in more trying circumstances. Not all the development work had been wasted, though, and lessons learned during the production of this V12 eventually trickled down into the road-going versions.

As in so many Jaguar launches, the publicity team liked to lean heavily on imagery from Jaguar's sporting heritage. A promotional film for the launch of the V12 E-type was organised to take place at MIRA in January 1971. Norman Dewis was tasked with driving the XJ13 down the straight at speed as part of the opening sequence. On the final lap, with the footage shot, a damaged wheel gave way on the banking, flipping the car end over end, before it rolled twice and came to rest on its wheels in the mud. Dewis, like all good racing drivers, let go of the steering wheel and also had the foresight to turn off the ignition during the accident. Luckily he emerged from the substantial wreckage unharmed, but the mangled car was then consigned to a corner of the Jaguar factory and put under a cover.

A few years later, Edward Loades, whose company Abbey Panels had played a part in creating the original bodywork for the car, spotted the crashed XJ13 in

Above: Not all the development work was wasted, as it gave Jaguar engineers valuable exerience with a V12 engine unit.

Right: Completing a demonstration run at the Goodwood Festival of Speed, the XJ13 was Jaguar's Le Mans car that never made it to the Sarthe.

the factory and offered to rebuild the car. It was to a specification 'similar' to the original, but with small variations to the original Malcolm Sayer design, such as the width of the wheelarches. It cost Jaguar £1,000 to reconstruct and was immediately put to work promoting the latest version of the XJ saloon. The company reportedly turned down an offer of £7 million in the late 1990s. Today, Malcolm Sayer's second-last design, that has a visual lineage to the C-X75, can be seen at the Jaguar Heritage Trust Collection at Gaydon in Warwickshire.

JAGUAR
4868 WK

and long-since dispersed components. The restoration, which took more than 7000 man-hours, won CMC several awards, and became the basis for a limited edition of Lindner-Nöcker Low-Drag E-type replicas.

The success of the Lindner-Nöcker restoration was surely instrumental in triggering one final twist in the E-type Lightweight story. In 2014 Jaguar announced that it would be building the remaining six of the 18 Lightweights originally planned, using the original chassis numbers that had been allocated on a dusty old Browns Lane register in the early 1960s.

Original plans were retrieved, old bonnet presses located throughout England and the cars built to original factory specifications. Sales were co-ordinated by former *Autosport* magazine director Tony Schulp, with each of the six offered to an enthusiast with a genuine interest in racing them, rather than to a collector or a car museum. The success of the programme, in which the Jaguar engineering team re-learned the company's former skills of hand building, would be used again in 2016.

JAGUAR E-TYPE LIGHTWEIGHT

ENGINE inline six-cylinder

CAPACITY 3781cc

BORE X STROKE 87 x 106mm

COMPRESSION RATIO 9:1

POWER 344bhp

VALVE GEAR dual overhead camshafts

FUEL SYSTEM triple HD8 SU carburettors

TRANSMISSION four-speed manual

FRONT SUSPENSION independent wishbones, torsion bars, anti-roll bars

REAR SUSPENSION independent lower wishbone, upper driveshaft link, radial arms, twin coil springs

BRAKES 4-wheel disc

WHEELS centre-lock peg-drive

WEIGHT 2484 lbs (1127 kg)

MAXIMUM SPEED 165mph (266kmh)

PRODUCTION 12 in 1963 and 6 in 2014

Left: In 2014 Jaguar completed the production run it had begun in the 1960s by using the final six chassis numbers it had originally issued.

JAGUAR E-TYPE SERIES 2

Although there had been certain changes to the 4.2-litre E-type that had given it the unofficial title 'the Series 1½' car, in October 1968 the E-type was given a substantive revision. This was prompted mainly by forthcoming safety and emission legislation in the U.S. but also by the experience accumulated from seven years of production.

The Series 2 car was visually distinct from its predecessor, having a larger 'mouth' as part of significant improvements to the cooling system. The grille was enlarged by around 70%, and behind it a bigger radiator and two battery-powered cooling fans had been added which all helped to keep the running temperature down.

The brakes, crucial to a car with the E-type's performance, were also enhanced, and now came with larger calipers front and back. When it came to going rather than stopping, the first gear ratio was lowered to make setting off easier. Power steering was now an option for the first time.

There were changes to the exterior lighting. The

Left and Below: The 'Series 1½' car had a number of safety revisions introduced for the American market.

Series 2 dispensed with the Series 1's headlamp coverings, which had significantly dimmed the beams, and instead the chrome trim around them was revised, being now much wider above the lamps. The indicator units were enlarged front and back, and at both ends they were now positioned below the bumpers.

The bumpers themselves were reworked. The front bumper was raised to just below the headlamp fairing and now incorporated the chrome splitter across the grille. In the centre of the grille it held an oval version of the 'growler' badge where the Series 1 badge was circular. At the rear, the bumper remained level with the crest of the rear wheel arch, but was now one continuous piece; the number plate, which had been mounted in the gap on Series 1, now sat below the bumper, and the twin exhausts sat either side of the plate – they had been side by side in the middle of the Series 1's rear. Surprisingly, the front number plate continued to be pasted onto the front of the bonnet, almost as an afterthought.

Under the bonnet, British versions of the Series 2 still used triple SU carburettors, but in the U.S. the car was now supplied with twin Strombergs, a de-tuning exercise to comply with American emissions legislation. New North American regulations for the automotive industry were responsible for several changes in the E-type – the Strombergs, for example, and a system for recycling gases back into the engine rather than releasing them into the atmosphere. Although these amendments (first introduced in the America-only Series 1½) were legally essential and environmentally beneficial, many E-type enthusiasts felt that the Series 2's performance was weaker as a result.

Above: The dashboard switchgear was radically overhauled from the original model, and now the E-type had recessed door handles.

American regulations also required the change inside from toggle switches to rockers on the dashboard. Elsewhere the interior showed only minor changes – revised, better padded trim on the doors, reclining seats with 'breathable' perforated leather upholstery, a cigarette lighter in the dash, and the horn on a stalk on the steering column rather than in the centre of the steering wheel. Air conditioning was introduced as a new option, for the North American market only.

The Series 2 came in all three styles – two-seater

coupé, roadster and 2+2 – and was produced from August 1968 to October 1970 (July 1970 for the 2+2). In its relatively short lifetime over 18,000 were sold, of which fewer than 3000 were right-hand drive. Although the roadster far outsold the coupé and the 2+2 in the States, in the UK the reverse was true: the roadster was the least popular model. Overall, in its three years of production the Series 2 sold more than the Series 3, which had a four-year production run; and the Series 2 sold almost half as many as the Series 1 did in its original eight-year run. Not bad for the often overlooked sibling in the E-type family.

Above: Apart from changes to the bumpers and headlights, the Series 2 had considerably improved brakes with three-pot calipers at the front.

ARIZONA
GRAND CANYON STATE

JAGUAR E-TYPE 2+2

The phenomenal success of the E-type after its launch in 1961 caught Jaguar by surprise. For the next five years it was all the company could do to keep up with the flood of orders. While it caught its breath and adjusted to the unexpected popularity of the new car, only a few cosmetic changes were made to the original specifications.

The first major event in the E-type's life was the introduction in 1966 of a 2+2 version of the coupé. Behind the two front seats a shallow upholstered bench seat could accommodate two small passengers or hand luggage. The new version never challenged the roadster or fixed-head cars for sales, but was popular enough to be offered in the E-type's Series 2 and Series 3 incarnations as well. When the Series 3 was launched, the coupé was *only* available as a 2+2.

The 2+2 was housed in a longer body over a longer wheelbase. The extra nine inches made room for wider doors. A broad strip of chrome trim ran across the doors at waist height and on to the rear wing, an identifying feature of the 2+2, apart from its sheer size. Under the shell there was now enough space for a Borg-Warner automatic transmission, which was an option that the American market had been crying out for.

The windscreen was more steeply set to extend the space in front of the driver and passenger's faces. In the Series 2 edition of the 2+2 it was enlarged and moved even further forward, almost to the edge of the bonnet, doing away with most of the scuttle. Overhead, there was an extra three inches of headroom, and there was a sense of much more room inside. The extra length of the body went some way to balancing the new, larger fixed head, but the 2+2 couldn't help looking a little top-heavy. Larger all round, it was nearly 100kg heavier than the two-seater, but could still manage nearly 140mph.

Access to the back was by tilting the front bucket seat backs forward and squeezing through. The backs were released by a lever which was only operable from the front. So rear passengers depended on those in front to let them out until, with Series 3, the lever was extended to be reachable from behind.

The upper half of the rear backrest folded forward to increase the capacity of the boot in the absence of passengers – the two-seater already had such a facility for the partition between boot deck and cockpit.

Below: The 2+2 made its debut at the New York Motor Show in 1966. The fixed-head coupé was the only model to offer automatic transmission.

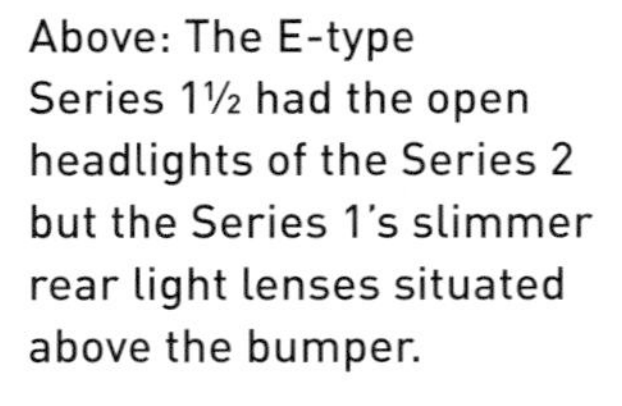

Above: The E-type Series 1½ had the open headlights of the Series 2 but the Series 1's slimmer rear light lenses situated above the bumper.

Right: The E-type is most associated with wire wheels, but both wire and factory-fitted solid wheels were available.

Left: The 2+2 was available in Series 1, 2 and 3. The Series 3 was the biggest seller with 7,297 sold, although the fixed-head coupé wasn't available in anything except 2+2 for Series 3.

JAGUAR E-TYPE SERIES 3

The success of the XJ6 saloon in the important U.S. market was perceived by Jaguar as the first step towards a multi-cylinder engine version of that car. Much as they had used the Jaguar XK120 to give the new XK engine unit a low-volume trial, so their new 5.3-litre V12 would be first installed in the latest iteration of the E-type, the Series 3.

The new 5343cc all-aluminium V12 was hailed by some as the best production car engine in the world. Jaguar's first attempt at a V12 had grown from the XK unit and was a 5-litre racing-type four-cam that proved to be too bulky for production use (though it was used in the XJ13 project)

The new 5.3-litre, 90 x 70mm engine was largely the brainchild of Wally Hassan and featured flat cylinder heads with combustion chambers formed in the piston crowns. Even with four Stromberg carburettors it produced 270bhp and gave the E-type back the performance it had lost over the years as it gained weight.

It was a bold automotive statement. At the time, the V12 engine was the only one in mass production and the first to be manufactured in volume since the Lincoln engine of 1948. The new E-type Series 3 was launched at Palm Beach, scene of many Jaguar sporting triumphs, in March 1971. Production of the Series 2 had stopped a full five months earlier, and the motoring press were eager to see what new life could be breathed into the ten-year-old E-type.

The new car had filled out a little compared to previous versions, but was every inch an E-type from its long nose to its generous rear. It used the longer chassis of earlier 2+2 versions, and had the luxury of wider doors which that chassis allowed. The boot was big too, whether sloping under the hatchback of the coupé or tapering more gently from the folded hood of the roadster.

The track was widened to accommodate the larger engine, and wider tyres now sat under flared wheel arches, contributing to the stability of the powerful ride. Pressed steel wheels were standard issue now, with the wire wheels of both earlier series an alternative option. Both wheels were also available in chrome.

To feed the increased air intake of the V12, the Series 3 had a wider grille, with a cross-hatch pattern of chrome-plated vertical and horizontal lines. At the other end of the car, exhaust gas was expelled via four pipes configured in a fishtail arrangement. In 1973 these were reduced to two. The bumper overriders were modified to absorb impacts, in line with incoming American safety legislation.

All these little changes – the arches, the wheels, the grille, the exhausts – were certainly benefits in themselves, although taken together they conspired to detract from the classic lines of the E-type. But under the bonnet, there was no mistaking the continuing authority of the Jaguar engineering. The V12 was barely heavier than the XK engine because its huge block was

Right: The Series 3 roadster, while remaining a two-seater was based on the 2+2 wheelbase.

Below left: Just as the XK engine had been trialled in a sports car before its ultimate destination, a high-volume saloon, so the new 5.3-litre V12 engine got a try-out in the Series 3 E-type.

MJH 26K

JAGUAR E-TYPE SERIES 3

ENGINE V12

CAPACITY 5343cc

BORE X STROKE 90 x 70mm

COMPRESSION RATIO 9:1

POWER 272bhp

VALVE GEAR two single overhead camshafts

FUEL SYSTEM 4 Zenith carburettors

TRANSMISSION 4-speed all-synchro or 3-speed automatic

FRONT SUSPENSION independent wishbones, torsion bar

REAR SUSPENSION independent lower wishbone, upper driveshaft link, radial arms, coil springs

BRAKES 4-wheel disc

WHEELS pressed steel, bolt-on

WEIGHT 3226 lbs (1463 kg)

MAXIMUM SPEED 146mph (235kmh)

PRODUCTION 15,292, 1971-1974

of alloy rather than iron. The comparability in weight gave the Series 3 similar handling characteristics to those of its XK-powered predecessor, and comparable top speeds.

The Series 3 could go from 0 to 60mph in under seven seconds, and new power steering helped the driver to handle the pace. The power steering also allowed Jaguar to replace the big old wooden steering wheel with a smaller one in leather trim.

It was easy to lose revs when changing gear, but the V12 delivered such high torque that drivers didn't have to shift much: first and fourth gear, which drivers could maintain comfortably at as little as 10mph, were all that were needed! Flat out, though, drivers felt the lack of a fifth gear, even more so with the three-speed automatic option. That was a comment on the power of the engine as much as on the well-spaced ratio of the gears.

Inside, the use of the longer wheelbase meant that all Series 3 coupés came as 2+2s, with the shallow rear seats for shopping or small children – although in fact a redesign of the seating did manage to squeeze a few more inches of legroom both front and back. The roadster was supplied only as a two-seater. Like the E-types of earlier series it had an optional saloon-style hardtop, with a more abruptly sloping rear window

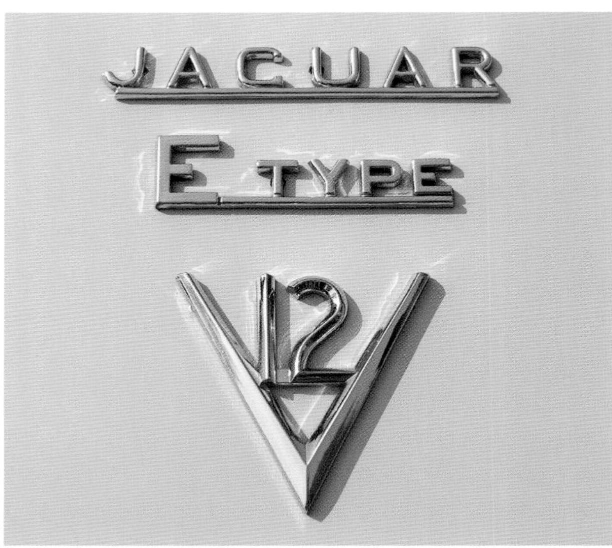

than the coupé's hatchback, more of a bubble than a wedge, to preserve access to the boot lid.

This latest version of the E-type was well- but not rapturously received. It sold steadily but not quickly; and some dealers, anticipating a greater volume of sales, found themselves overstocked. E-type waiting lists were suddenly a thing of the past, and Jaguar was in the dispiriting position of having to advertise the car extensively for the first time.

An XK-powered version of the Series 3 was announced at the launch of the V12 machine; but although a few were built, it never went into full-scale

production. American emissions legislation was having an impact on engine design, threatening the viability of old-timers like the XK.

To add to the company's woes, a further batch of American regulations was on its way, addressing issues of safety in the event of crashes, which it was clear the E-type could not meet. The writing was on the wall for the car, and Jaguar began to focus on the future in the form of the XJ-S. Regular production of the Series 3 coupé ended in 1973. The roadster continued for another year or so, to the end of 1974.

Sir William Lyons retired as chairman, his last formal connection with Jaguar, in 1972. It really was the end of an era. To mark its passing, in late 1974, Jaguar produced one last batch of E-types, a special edition of 50 Series 3 convertibles. They came with chrome wire wheels and a special plaque on the dashboard bearing the chassis number and the signature of Sir William, for whom the E-type had been his favourite creation.

All but one of the edition were painted black, as if in mourning, and the very last car off the line was kept by Jaguar for its collection. The second last, the last new Jaguar E-type to be sold in Britain until the six Lightweights of 2014, was painted British Racing Green. Despite the significance of these valedictory models, the last E-types proved difficult to sell, and many were only acquired at reduced prices. The sports car-buying public had moved on.

The bare figures show that the Series 3 sold reasonably well, albeit with a push from the marketing department: well over 7000 of each model was built. By comparison, of all the ten different versions produced in the E-type's 13-year history, only the two earlier roadsters sold more than 8000.

Above: After ten years in production, interest in the E-type tailed off and Jaguar were forced to devise marketing schemes to keep the cars moving through showrooms until the arrival of the XJ-S.

JAGUAR MARK X TO XJ12

As the 1960s dawned, Jaguar was gearing up for the launch of a major new saloon car. The independent rear suspension that had been so successfully used in the E-type had been developed for what the company considered its major market, a big saloon car for the U.S. and the major new model was the Mark X (Mark 10). From its inception, the E-type Jaguar had been viewed by Sir Williams Lyons only as a low-volume sideline.

The Mark X was the largest Jaguar yet produced, with a unitary chassis construction and highly distinctive, rounded lines that showed a design progression from earlier models (and would influence the future XJ6) together with a radiator grille and front panel that projected forward at the top. The amount of leg room inside was enormous. The Mark X used the 3.8-litre triple carburettor E-type engine, mated to either a four-speed gearbox or a Borg-Warner automatic. For a large car it had a startling turn of speed, it could manage 122mph and a 0-60mph time of 10.4 seconds. The new suspension provided remarkable handling qualities too, if the driver and passengers were not put off by a fair degree of roll.

It was a remarkably refined and sophisticated car for its day, with ride qualities not exceeded by any other luxury saloon on the American market, yet it failed to sell in great numbers. It was considered to lack the grace of the previous saloons such as the Mark VII and the early cars suffered from cooling system and electrical problems. And while Jaguar sold 47,200 of the Mark VII, VIII and IX in the decade from 1951 to 1961, they managed to sell only 24,000 of the Mark X

Below: The Mark X was a particularly refined car for its day, but it proved to be one of the few cars where Sir William Lyons' sales projections were optimistic. The important U.S. market thought it lacked grace, despite the space.

Left and Below: The 240 and 340 saloons were introduced in September 1967 to provide a cheaper entry-level car than the Mark 2.

and its successor the 420G up to 1970.

Jaguar under Lyons had always achieved much with the minimum of resources thanks to engineering ingenuity and careful investment. During the middle and late 1960s a number of new and revised models appeared based on existing components and engineering first developed for the Mark X. The first was the original S-type, which combined a stretched Mark 2 bodyshell with the Mark X's independent rear suspension. It retained the Mark 2 doors and roofline, but had extended rear wings, a boot compartment and a remodelled front end. Buyers could choose a 3.4- or 3.8-litre engine. It was more comfortable than the Mark 2, yet cheaper and more compact than the Mark X.

In October 1964 the trusty XK engine capacity was enlarged again, this time to produce a 4.2-litre unit aimed at providing more power for both the E-type and the Mark X. Then, in 1966, the company combined the 4.2-litre engine with the S-type saloon body and grafted on the front end of the Mark X to produce the 420. The company hadn't finished with the Mark X, though, and restyled the interior to create the 420G (the G was for 'Grand').

The final stage of getting the most from the Mark 2 tooling came in September 1967 when the 'budget'

Below: The S-type was a modified Mark 2 bodyshell with the Mark X's independent rear suspension. It was visually similar to the Mark 2.

Left and Below: In 1966 the 4.2-litre Mark X was upgraded to the 420G (G for Grand) with a restyled interior.

240 and 340 saloons were introduced. Sales of the Mark 2 had held up surprisingly well through the 1960s and these cars were lightly modernized, slim-bumpered versions of the Mark 2 with 2.4- and 3.4-litre engines that had been given the straight port 'gold top' cylinder head. Their role was to provide a cheaper, entry-level car as a standby, in case of problems with the production or popularity of the brand new XJ6 scheduled for 1968.

Ever since the early 1960s, work had been underway on a Mark X replacement. The new model needed to be shorter, narrower, lower, and ideally quieter, faster and roll less. The new car would be the XJ6 and represented Sir William Lyons' apotheosis in car development. It was quieter and had a better ride than a Rolls Royce, it out-handled a Mercedes, it was faster than a Cadillac (important for the U.S. market) and it looked more graceful and modern than any of them.

The car used the accumulated knowledge of automobile refinement that the Jaguar factory under Bill Heynes and Bob Knight had acquired since the early 1950s. However the car's basic engineering was not radical, it employed the same engine, transmission

and suspension systems that had been proven in the Mark X. It was available with a 2.8-litre version of the XK engine, favoured by European markets which heavily taxed engines with a capacity over three litres, while in the States the 4.2-litre version was the more popular. Dunlop had produced a new, wide (205 section) low-profile radial tyre especially for Jaguar which gave the car supreme grip for its size.

The XJ6 was viewed by Jaguar as a first step. The ambition was to harness the power of a V12 engine in the chassis. Just as the XK120 had been used to give the new XK engine a trial run, so the V12 was developed and installed in the Series 3 E-type before it went near a saloon. The 5343cc all-aluminium

Above and Right: The XJ6 introduced in 1968 was considered to be Sir William Lyons' crowning glory, combining all the automotive lessons learned on the previous saloons, particularly the Mark X.

Left: It was powered by the 4.2-litre XK engine used in the 420G, although a 2.8-litre version was available in European markets.

Right: The Series 2 XJ included a two-door version. Although the Series 2 was launched in 1973, problems with the pillarless construction and the window seals delayed it until 1975.

V12 was a major step forward; equipped with four Stromberg carburettors it produced 270bhp, but made the new XJ12 particularly thirsty, returning 12mpg.

The XJ12 was launched in August 1972 and was followed up in August by a longer wheelbase version. The extra four inches were available from September 1973 with the arrival of the Series 2 range of saloons. A two-door pillarless coupé on the original wheelbase, complete with vinyl roof, was also announced, with production commencing in April 1975.

The Series 3 XJ range emerged in 1979 with a new roofline and facelifted interior. There was an optional five-speed manual gearbox, and the fuel-injected 4.2-litre engine first used in late-model Series 2 cars in North America was adopted as standard, along with the option of the normally carburetted 3.4-litre unit which had first appeared in 1975.

By the late 1970s the XJ6 should already have been succeeded by a new saloon coded the XJ40. However the interim Series 3 XJ6 proved so successful that the XJ40, which at the same time was proving troublesome, was put into further development until its launch in October 1986 in Europe and April 1987 in New York. On a component by component basis, the new car had little in common with its predecessor, and there were virtually no carry-over parts from the XJ6 Series 3 to the XJ40. Apart from the name. Such was the esteem in which 'XJ6' was now held that the new car retained the important branding.

The 'XJ40' had a new, lighter, multi-link, 'compliant' rear suspension and refined the Series 3 front suspension. The new family of AJ6 engines gave the lighter XJ40 greater performance than the Series 3. But despite spending a long time in development and becoming the most tested saloon Jaguar ever put on the road, there were many initial faults and it would take over a year before the car achieved Series 3 reliability.

The XJ40/XJ6 received its first upgrade in September 1989 when the twin-cam 3.6-litre engine was increased to 3980cc. The new unit moved the power output up from 199bhp to 223bhp. It was discovered that Jaguar owners preferred conventional instrument dials to the hi-tec fluorescent display that had been offered before

Above: The Series 3 XJ first seen in 1979 had restyling courtesy of Pininfarina and was a sales success.

Right: The styling cues that had begun with the original XJ6 from 1968 were continued through to the X308-generation of XJ8.

and so these were now returned to the dashboard.

Due to a rise in its popularity, the Series 3 XJ6 didn't disappear from showrooms in 1986, but even after it did, the Series 3 XJ12 continued in production. It sold 300-400 units a year and in Japan it was revered as a 'living classic'. Finally, on 30 November 1992, the very last Series 3 XJ12 rolled off the production line at Browns Lane. It was the final car of the original XJ breed. Since its first assembly in 1968, some 177,240 cars had been made. The XJ12 badge would continue, but from 1993 it would be attached to the XJ40/XJ6 bodyshell.

Above: The XJ40 body shape was an evolution of the Series 3 but there were few carry-over parts from the previous car.

JAGUAR XJ-S

Jaguar in the mid-1970s had a problem: how to follow the E-type. Jaguar knew that the icon's successor had to be a success. The company needed a hit; it was in difficulty, racked by industrial disputes and with a growing reputation for producing attractive but poorly built, unreliable cars. Work had begun on the E-type replacement in 1965, ten years before the eventual unveiling of its successor, and the design went through many conceptual changes before the final clay model was produced.

Jaguar enthusiasts and the press were looking forward to the company's solution to its problems, but when the XJ-S first appeared before the public at the Frankfurt Motor Show in 1975, the response was one of disappointment. This was not what they had been expecting. There were no graceful E-type curves, but a boxy, angular body that shared none of the design characteristics of the previous car. Journalists speculated that this would be the shortest-lived Jaguar in the company's history. As it turned out, Jaguar had produced a car which would go on to outsell the E-Type by far; the XJ-S became one of the most successful cars that Jaguar ever built.

One of the aspects of the new car's design which disconcerted critics was the presence of 'flying buttresses', stretching from the corner pillars of the rear passenger windows to the boot lid. The feature was a hangover from an earlier design for a mid-engined sports car (as in the XJ13), where mid-chassis structural stability was crucial. Over time the XJ-S morphed from the original mid-engine concept to a front-engine luxury grand tourer, designed to conquer the autobahns and Interstates rather than the B-roads. Somehow the flying buttresses remained after the change of focus and (again with serendipitous hindsight) became one of the defining design elements of the XJ-S.

Jaguar had spent years conceiving and designing

Above left: The Jaguar XJ-S was the last V12 Jaguar to be offered with a manual transmission, though very few took up the option. The car above has the High Efficiency engine introduced in 1981.

Above: The XJ-S launched in 1975 was the last car from D- and E-type designer Malcolm Sayer and it was unlike anything Jaguar had yet produced, a Grand Tourer.

the XJ-S, and came to the conclusion that two-seater sports cars were a shrinking market. The E-type had a loyal and enthusiastic following. But by the end of its life, it was being overtaken by rivals of more advanced design. And so Jaguar decided that it would be folly simply to create another E-type. Instead it fixed its sights firmly on the luxury tourer market already occupied by Aston Martin, Porsche and Ferrari. Jaguar, in declining financial health, badly needed the kind of profit that this lucrative market generated.

Although an open version was originally planned, due to proposed (but unimplemented) American legislation that would outlaw convertibles, the XJ-S was launched in coupé form only.

Compared to the E-type, the XJ-S was a bold and ambitious statement. The radical design was begun by Malcolm Sayer and completed after Sayer's premature death at the age of 54 in 1970. The car may have confused the public, but what rolled off the production line was a very long, low, comfortable cruiser powered by a proven V12 engine. It made a clean break with what had gone before.

In 1975 Jaguar was a brand still best known for its racing pedigree. With the XJ-S, the company presented a car which they hoped would reposition them as an affordable luxury marque, generating volume sales on

Above: Jaguar enthusiasts had to go to specialist customizers for the first soft-top versions. This is a rare Lynx Spider, one of 180 made.

Left: The original 5.3-litre V12 produced 285bhp but returned a poor 13mpg. The later HE version increased this to 16mpg while power was increased to 299bhp.

JAGUAR XJ-S

ENGINE V12

CAPACITY 5343cc

BORE X STROKE 90 x 70mm

COMPRESSION RATIO 9:1

POWER 285bhp

VALVE GEAR two single overhead camshafts

FUEL SYSTEM injection

TRANSMISSION 3-speed auto, 4-speed synchro

FRONT SUSPENSION independent wishbone, coils, anti-dive

REAR SUSPENSION independent lower wishbone, upper drivelink, radius arms, coil springs

BRAKES 4-wheel disc, rear inboard

WHEELS alloy

WEIGHT 3718 lbs (1686 kg)

MAXIMUM SPEED 153mph (246kmh)

PRODUCTION 115,330, 1975-1996

both sides of the Atlantic. At an entry price of £8900, it also undercut their competitors by some margin.

If the public were at first disappointed by the XJ-S, so must Jaguar have been. Sales were very sluggish. A production figure of only 1245 cars in the first 12 months was bad enough but worse was to follow. Sales declined over the next few years at such a rate that the production of the XJ-S came very close to being shelved.

A global fuel crisis was partly to blame for the continuing lack of demand, and by 1982, specifically for the larger U.S. market, Jaguar introduced the HE (High Efficiency) version of the 5.3-litre V12, with a redesigned cylinder head and a much higher compression ratio of 10.5:1 for improved economy and power. Safety and emission regulations, especially in America, were a crucial influence on the evolving design of the XJ-S as they had been on the E-type. John Egan was now in charge of the company and under his stewardship Jaguar revamped the assembly line and the

Above: Jaguar's own XJ-S convertible made its debut at the 1988 Geneva Motor Show. Jointly developed by Karmann in Germany, there were 130 new or modified panels.

Above: The XJ-SC was a stop-gap model until Jaguar could introduce a true factory-produced roadster from 1988.

build processes, resulting in higher assembly quality. Higher quality was also reflected in new interior trim, which re-introduced Jaguar's hallmark walnut burr veneer on the fascia and door trim panels.

At the same time Jaguar conducted a searching review of both the car and its advertising strategy in the hope of turning the sales curve around. The company recognised that the market for a luxury V12, although profitable, was quite small, and put them in direct competition with Lamborghini and Ferrari. The solution was a sideways move, in 1983, to a new version with a smaller 3.6-litre straight-six engine, that necessitated the introduction of a 'power bulge' in the bonnet to accommodate the vertical engine.

Initially, the 3.6-litre model was only available with manual transmission, but customers expected that a grand tourer, even one with a reduced engine capacity, would have an automatic option. The automatic version of the 3.6-litre XJ-S duly entered the market in 1987, and was rewarded with strong sales.

Before 1983 there was still no option for open-top motoring in the XJ-S, a disappointment for Jaguar purists for whom the company's long tradition of iconic soft-top roadsters was its main attraction. As

noted earlier, Jaguar had not planned to introduce a convertible, but strong customer demand from the States made it inevitable.

In 1983, a short-term solution was adopted in the form of the XJ-SC, a cabriolet built by Tickford Coachworks on completed coupé bodyshells, with a manual fold-down roof and twin-targa panels, which could be stored in the boot. The rear two seats were sacrificed to allow interior luggage space, but the rear windows remained, with a targa bar across the roof. The resulting XJ-SC may not have been a full convertible, because of its targa top, but it delivered open-top motoring at a fraction of the cost of a full convertible. A V12 version of the XJ-SC arrived in 1985.

A full convertible built by U.S. coachbuilding, company Hess & Eisenhardt, had been originally offered though American dealers. This was a high-quality conversion, with modifications to the fuel tank to allow the folding roof to retract fully, and with additional structural elements to ensure the rigidity of the car. Available from 1986, the number of commissions for these expensive convertibles (approaching 900) left Jaguar in no doubt that a production soft-top would be a valuable boost to the sales, and therefore the lifespan of the XJ-S.

These rather *ad hoc*, unsatisfactory fixes for the convertible problem, sanctioned by Jaguar, did boost sales, although they were not the sort of elegant design solution to which Jaguar enthusiasts were accustomed.

The improvised arrangements did however prove that there was a need for an official XJ-S convertible. From 1988 Jaguar finally decided to offer a production soft-top incorporating a fully-retractable, electrically operated hood with a heated rear glass screen. Their direct competitor in the class, Mercedes, already offered such a solution in the highly regarded SL series.

Jaguar was able to learn from the earlier experiences of Hess & Eisenhardt, and the new model included a strengthened floorpan and improved rigidity. The

G14 XJS

Above: Le Mans winner Martin Brundle at the wheel of the TWR Jaguar XJ-S as it muscles its way round the hillclimb course at the 2013 Goodwood Festival of Speed.

convertible was an immediate hit, even with a 20% price hike over the coupé, and it remained the most popular version of the XJ-S to the end of its production.

Although an unlikely competitor on the track, Bob Tullius and his Group 44 team had competed the car successfully in the 1977 and 1978 Trans-Am series in the States. Tom Walkinshaw and his racing outfit TWR (Tom Walkinshaw Racing) also campaigned the car in the European Touring Car Championship between 1982 and 1984. They won the European competition outright in 1984, which encouraged John Egan to pursue the ultimate goal and try to regain Le Mans glory for the marque.

Tullius's Group 44 team got the original contract to challenge in the 24 Heures Du Mans, but when that bid faltered, the more experienced Walkinshaw was given the job of capturing the World Sports Car Championship and the ultimate prize, Le Mans victory. The team came close in 1987 when they dominated the world series, but failed in the ultimate sports car endurance race. A year later, the TWR-run Jaguar XJR-9LM of Jan Lammers, Johnny Dumfries and Andy Wallace took the top prize

On the back of this sporting success, in 1988 Jaguar and TWR created a new company, JaguarSport, to build sportier versions of its cars. The resultant XJR-S (and XJR) was big on performance, style and comfort. A manual gearbox was again offered on the V12, and various trim and engine upgrade options, including an enhanced body kit, suspension improvements and exclusive alloy wheels. Over 800 of the XJR-S were produced for the world market including a 6.0-litre version (787 coupés and 50 convertibles for America) and it would stay on sale until 1993. The highly specified models played their part in further restoring the reputation of the company.

Facelift Model

Following the cancellation of the XJ41 sportscar project by Jaguar's new owner, Ford, the company knew it had to extend the life of the XJ-S until a new car could be developed. A makeover was required to maintain its place in the market and to suit the new decade. By 1992, the redesign was complete and the name was subtly altered: the old XJ-S was to become the 'new XJS' (without hyphen). The famous buttresses stayed, but the rear windows were enlarged, and the XJS now sported softer, more aerodynamic lines, with optional body kit.

The V12 version boasted a new power bulge on the bonnet, and both cars had revised rear-end profiles and lights, with body-coloured bumpers and uprated headlamps. For the convertible, a new subframe was designed. The manufacturing process was also comprehensively revised, with additional rustproofing and further attention to build quality. Inside the cockpit, the fascia was all-new, with traditional dials and improved trim. The result went down well with the motoring press and public. The XJS shook off its angular look, nearly 20 years old now, and was transformed into a smooth luxury tourer for the 1990s.

The most dramatic changes in the new XJS took place under the bonnet. For the new model, Jaguar had considered cutting down the existing V12 powertrain to a V8 or V6. In the end they decided to upgrade the existing V12 to six litres, with major improvements in efficiency and power output, and attached the uprated engine to a new GM 4L8 transmission. The 3.6-litre engine was replaced with an entirely new machine, the 4.0-litre AJ6 engine, which more than delivered in power output.

To give some idea of the longevity of the outgoing V12 5.3-litre and the straight-six 3.6-litre, the new AJ6 was only the third new engine that Jaguar had ever produced from scratch. As with any new engine, the AJ6 underwent a series of refinements in the first few years, including coil-on-plug ignition, which lifted the power output to 238bhp. The new 4.0-litre became the XJS of choice, and easily outsold the more powerful 6.0-litre V12.

Jaguar also addressed the lack of rear seats in their most popular model, the convertible, and added a new 2+2 configuration, which involved a major redesign of the floorpan and rear chassis structure. The 4.0-litre coupé and the 2+2 received improved rear brakes, with new outboard rear disc brakes, to replace the ageing and complex inboard brakes used on earlier models.

The cosmetic and engineering upgrades delivered

Above: Lynx Engineering were involved in many Jaguar-based projects including the production of replica D-types and the XKSS. The development of a sporting brake, the XJ-S Eventer, was their most ambitious project.

Right: Remarkably the Eventer was lighter than the production coupé. Each car was built to order and Lynx produced 67 in total over a period of 16 years.

a more luxurious car with greatly improved handling. Although strictly speaking it was more of a grand tourer than a sports car, the improvements in handling made for a more responsive driving experience, whether on motorways or on twisting back roads. In its final incarnation, the XJS became a desirable, glamorous and now highly driveable machine.

After a shaky start, the XJ-S/XJS is now credited with turning Jaguar around at a critical moment in its history. The XJ-S confounded its critics, overcoming slow sales in 1975 that threatened the entire project. Although originally considered an unworthy replacement for the iconic E-type, it achieved a much longer production run than its famous predecessor and outsold it by over 43,000 cars. The last of over 115,000 XJS cars, a blue V12 coupé, was driven out of the factory on April 4th 1996, and the range entered the record book with the longest production run in the company's history: 21 years. By the end, the market had changed, and a smaller, sportier Jaguar was required. Work had begun on the XJS's successor, the XK8, a car ironically more in tune with the E-type.

Without the XJ-S, Jaguar might have disappeared like so many other British marques, ousted by Japanese and German manufacturers who brought better built, more reliable cars to the market. The XJ-S helped Jaguar to survive the uncertain days of the 1970s and rebuild its reputation around the world.

Right and Below: The facelift model of 1992 softened some of the angular lines of Malcolm Sayer's original design.

Left: The XJ-S/XJS had the longest production run of any Jaguar car from 1975 to 1996.

Left: The 6.0-litre XJR-S sports model. Through its lifetime the XJ-S had many short-run custom builds and limited edition models, such as the XJ-S Le Mans and the Celebration.

JAGUAR XJR-15

Jaguar's success in motorsport in the 1980s included, in 1988, the prize dearest to the company, the 24 Hours of Le Mans. The XJR-9LM's triumph in the contest was a major media story and the company was quick to make the most of its success. JaguarSport and their chosen partner TWR Special Vehicle Operations combined to produce one of the archetypal supercars of its time, the Jaguar XJR-15.

Built and designed by JaguarSport's factory in Bloxham, Oxfordshire, the company set out to build a ground-breaking vehicle; one which would draw on the engineering skills and experience gained at the trackside. The bar was set high from the start. It was to be a road-going supercar, with a chassis and bodywork entirely composed of carbon composites. With this ambitious choice of material, Jaguar was far ahead of other manufacturers. It would be another four years before McLaren adopted these techniques for the Gordon Murray-designed McLaren F1.

The XJR-15 was given a normally aspirated 6.0-litre V12 derived from the racing engine, with electronic fuel injection and fly-by-wire throttle, which delivered 450bhp. The engine was paired up with a five-speed manual gearbox, although, for the serious driver with racing intentions for his XJR-15, Jaguar offered the option of an unsynchronized racing box.

The XJR-9 racing chassis, a super-light monocoque construction of a carbon fibre and Kevlar composite, was adjusted for the XJR-15, with the lines pushed higher and wider, while retaining the squat racing stance. The use of a racing chassis for a road-going car meant that entry to the cockpit was a bit of a clamber: up over a wide sill, then awkwardly down into the low racing seats. But anyone attempting to squeeze their way inside was probably already aware that this was a street-legal racing car, not a leisure vehicle. Amongst the business-like curves of the composite body there was no place for a weekend suitcase.

The bucket seats held the occupants in a reclining position, which afforded an excellent view of the road ahead but not of the front of the car. This made driving at low speeds in everyday traffic situations perilous, but offered a superb seat-of-the-pants racing feel on the open road, and indeed on the track. The space in the cockpit was cramped, with sound insulation kept to the bare minimum. Although the driver and

JAGUAR XJR-15

ENGINE V12

CAPACITY 5993cc

BORE X STROKE 87 x 84mm

COMPRESSION RATIO 11:1

POWER 450bhp

VALVE GEAR twin overhead camshafts

FUEL SYSTEM electronic fuel injection

TRANSMISSION Eight-speed manual

FRONT SUSPENSION wishbones, inboard and horizontal pushrod-spring dampers

REAR SUSPENSION vertical coil springs

BRAKES 4-wheel discs

WHEELS alloy

WEIGHT 2315 lbs (1050 kg)

MAXIMUM SPEED 191mph (307kmh)

PRODUCTION 53, 1990-1992

passenger sat very close together, they were only able to communicate over the raucous V12 engine and thundering exhaust by a specially fitted headset and microphone communication system, more akin to driver and navigator in a rally car.

All cars were manufactured for right-hand drive, and in keeping with the XJR-15's racing pedigree, the gear lever was set to the right of the driver and very close to the steering wheel. The dashboard carried a minimum of controls and displays, another nod to its racing heritage. Racing-style analogue gauges were fitted in the instrument binnacle and four metal switches lay to the right, including a pump switch and engine cut-off. The dash was constructed of bare carbon fibre, with none of the leather or veneer embellishments associated with normal road cars. This functional interior aesthetic by Peter Stevens, in a car designed emphatically for performance, was admired by style reviewers as being entirely appropriate. Stevens, it should be noted, went on to style the McLaren F1.

The aerodynamic body design, a graceful, subtle blend of power and beauty by Tony Southgate, reflected Jaguar's constant mantra of pace and grace (but not, in this instance, space). The front splitter sat low, racing-style, below a sculpted and curving air intake which rose up the short bonnet to the domed cockpit. At the rear, the long, louvred engine cover drew the delicate lines of the car towards a very practical rear spoiler.

It may have had the pedigree, the engine and the body of a racer, but Jaguar did have to make some compromises if they were going to market it as a road-going vehicle. While retaining the flat race-regulation

floor and the huge venturi tunnels which maximized downforce, the ride height was lifted to allow use on public roads. The fully independent front suspension was made up of wishbones connected to pushrods and dampers, allowing the suspension configuration to be mounted horizontally under the bonnet. At the rear, both the V12 and the transmission were bolted to the stressed bulkhead, and the coil springs were vertically mounted in uprights within the rear wheels. Using non-adjustable Bilstein shock absorbers, the suspension was softened to a degree sufficient to tame the wildness of the racing ride, and huge four-pot calipers were fitted to the vented and crossed steel disc brakes.

The result of Jaguar's determination to bring their racing pedigree to the road was a head-turning supercar, with a phenomenal time from 0 to 60mph of 3.9 seconds, and a potential top speed of over 190mph. From the launch at Silverstone in 1991

Previous page: Thierry Tassin, with a modified front bumper, leads Will Hoy and Bob Wollek at Spa-Francorchamps in the finale of the Jaguar Sport International Challenge in 1991.

Left: The XJR-15 was the first road-going car with a bodyshell composed entirely of carbon and Kevlar composites.

Below: With a production run limited to 53, the XJR-15 was similar to the original concept of the XKSS in the 1950s.

the XJR-15 gained considerable attention, as much because of its price tag of just under $1 million (then worth around £500,000, a dizzying amount) as the fact it was the world's first ever carbon fibre road car.

In 1991, 16 of them raced in the single-make Jaguar Intercontinental Challenge series which supported three Formula One races at Monaco, Silverstone and Spa. The hefty price tag for an XJR-15 included the preparation and maintenance of the car by JaguarSport during the series. Jaguar presented each of the winners of the first two rounds with an XJR-S road car, and gave the winner of the final race, German owner-driver Armin Hahne, $1 million in prize money.

Only 53 XJR-15s were ever built. Although some made it onto the open road, many were purchased by wealthy collectors and hidden away as investments, to reappear only at classic car shows, or in the international auction market.

In the XJR-15, Jaguar designed and delivered a near-perfect compromise between an aggressive racer and a road-going supercar. Critics agreed that the XJR-15 was more at home on the track than the road, but it could be driven home at the end of the day, and in some style. Thanks to the all-composite chassis construction, it was extremely light for a supercar, weighing little more than 1000kg, not far off the weight of an XJR-9. Despite the concessions made for road use, this two-seat supercar was the closest possible experience on the road to driving the track at Le Mans.

JAGUAR XJ220

The story of the XJ220, the first Jaguar production car to break the 500bhp barrier, is unique in the company's history. In 1987 Jim Randle, Jaguar's engineering director, began designing a road-going supercar which would also be able to race in the FIA Group B championship. The company set out to build a four-wheel drive, rear-wheel steer, race-capable road car. The XJ220 was to include a completely aluminium clad chassis with a built-in, FIA-approved roll cage, and a classic 6.2-litre V12 Jaguar powerplant. Variable aerodynamics would generate over 1300kg of downforce at 220mph.

The original design for the XJ220 emerged from the Saturday Club, an informal gathering of Jaguar's design team led by Jim Randle to discuss pet projects. At the time, a large proportion of the company's engineering development and design staff were committed to the redesign of the XJ-S and the new XJ saloon.

So Jaguar worked with other suppliers and engineering companies to make their vision for the Keith Helfet-styled supercar a reality. They made deals: if the suppliers designed and built the components to the required specification at their own costs, and they worked, then they would receive the contract from Jaguar for the parts, should the car go into production.

The top speed was reflected in the name of the new supercar, XJ220 recalling the similarly ground-breaking capabilities of the XK120, which claimed, then reached, a top speed of 120mph in 1949. From concept to unveiling, it was less than a year after the project began that Jaguar pulled the covers from the XJ220 prototype at the 1988 British Motor Show in Birmingham to a rapturous reception from press and public.

Jaguar announced its intention to build a limited edition of just 350. The price was yet to be determined,

Right: Built at the company's new Whitley engineering facility, the Jaguar XJ220 made its debut at the 1988 British Motor Show

Below: The Martini-liveried Jaguar XJ220 ran in the 1993 and 1994 Italian GT Championship.

but buyers could reserve one with a deposit of £50,000. Around 1400 potential customers came forward oversubscribing the offer by a factor of four.

Jaguar and TWR's joint company, JaguarSport was again tasked to build the XJ220 under a new company, Project XJ220 Ltd. TWR had great success and long experience of the racing V12 engines, and for the XJ220 they devised a reduced, 6.2-litre capacity version of the 7-litre V12 that won Le Mans in the XJR-9LM.

After the prototype's encouraging reception in Birmingham, the team had to consider how to turn the concept car into a production reality. Jaguar took a long hard look at its direct competitors in the market, Ferrari's F40 and the Porsche 959. Compared to the normally aspirated V12 which TWR proposed for the XJ220, both its rivals adopted smaller engines with forced induction. The Ferrari used a 2.9-litre twin-

turbo V8 producing 471bhp and the Porsche a twin-turbo flat-six delivering 440bhp. These more compact engines gave them the edge in the matter of power-to-weight ratio. A secondary issue for Jaguar's V12 was its prodigious fuel consumption, at a time when a global recession was beginning to bite.

In respect of this, Jaguar made the first of several significant departures from the original concept. To consternation from both the motoring press and prospective buyers who had put down a deposit based on the original specifications, the V12 was replaced by a turbocharged V6. Many of those who were treating the XJ220 as a financial investment, reacted by informing Jaguar that they would not be taking up their option to purchase, and demanding their deposit money back. As the 1990s dawned, it was a buyers' market for collectible cars.

Changing the engine was not a step lightly taken given the negative press and the financial backlash but it was the most practical option. Jaguar realised that the V12 engine would have great difficulty meeting existing and forthcoming emission regulations while providing the required power. The completely new 3.5-litre V6 engine, developed by TWR from the original specification of the Austin Rover V64V, was fitted with two Garret turbo-chargers. It was an impressive engine, developed for the Metro 6R4 rally car to compete in the spectacularly fast (but short-lived) Group B of the World Rally Championship. It was lightweight yet very powerful, with a 90° bank angle, four valves per cylinder and belt-driven, double overhead camshafts.

Above middle: Unlike previous Jaguars, the XJ220 had additional instruments in the driver's door.

Another factor in the decision was Jaguar's desire to give the XJ220 the capacity for four-wheel drive. Engine bay space was at a premium; by utilising the more compact V6, the front-wheel drive differential could be connected by a driveshaft through the V of the engine, an ingenious solution. But the idea became another casualty of cost-benefit analysis, along with the whole suggestion of four-wheel drive. Jaguar press releases assured the public that this was a sensible step. Rear-wheel drive, the company insisted, would be suitable for most occasions on the road or track; and the technical complexity of the four-wheel drive system would have a detrimental impact on the running costs for the customer. The original design was then adapted to rear-wheel only, with a limited slip differential connected to a five-speed gearbox, and synchromesh was fitted to the first two gears for ease of use. But for customers, it was another compromise of

the Birmingham Motor Show concept.

There remained plenty of areas in which the new design trumped the competition. The sparsely equipped Ferrari, for example, had exposed weld seams in the interior, like a race car, and no radio, electric windows or door handles. The XJ220 had a much more luxuriously appointed interior with Connolly leather trim, front- and rear-heated screens, air conditioning, a high-spec CD/stereo by Alpine and heated and electrically adjustable seats. The XJ220 still held to the original concept of electrically operated scissor doors, and a transparent engine cover to show off the motor, even if it was no longer a V12.

The original body for the concept car was hand-built over many hours by specialist craftsmen, and the swooping lines on the drawing board were challenging and difficult to realise. Aerodynamic efficiency was crucial, but could not include some of the more extreme designs adopted by race cars. The motorised rear wing of the concept, for example, did not make it to the production car. The venturi tunnels, in a modified form, did – a first in a production car, they were mounted low, just in front of the rear wheels.

Alongside the use of venturi, the XJ220 was one of the first production cars to harness underbody airflow. Through extensive use of a wind tunnel, Jaguar engineers shaped a car which, despite having road-car ground clearances, had a drag coefficient of 0.36 and a downforce of 3000lbs at over 200mph.

Ahead of the production car's release, former Le Mans winner Martin Brundle drove the XJ220 at the Nardo test track in Italy, a perfectly circular 12.5km track where supercars can be tested to their full potential. After Brundle had pushed the car to 212.3mph, Jaguar decided to adjust the rev limiter, and the XJ220 reached 217.1mph. With the variables of a circular test track taken into account, this equated to 220mph in a straight line. The XJ220 had fulfilled its promise.

In 1991 the first cars rolled off the production line in a purpose-built facility near Oxford. Many customers expressed disappointment not only with the modifications, but with a steep price rise, from a 1990 estimate of around £250,000 to a final £411,000. The launch quickly descended into bitterness and recrimination. Customers refused to take delivery, and some took Jaguar to court for failing to deliver on the promised concept. Others, ruined by the recession, were simply no longer able to pay such a high price for their indulgence.

The company offered customers the option to buy themselves out of their contracts, but the legal arguments rolled on for some time. Production ended in 1994, with only 281 cars built out of the projected total of 350. Despite a price drop to below £200,000, many of the remaining cars were left unsold, and a consignment of 150 cars was shipped to the U.S. for a single speculative buyer.

History has not been kind to the XJ220, but much of the criticism is undeserved. It was the right car, built at the wrong time, and in its day, the fastest car in the world. Between 1992 and 2000, it held the production car lap record at the Nürburgring with a time of 7min 46sec. It was a showpiece of engineering excellence. For all the litigation, the collapsing economy, disappearing customers and watered-down specifications, Jaguar did what it set out to do, produce a world-beating supercar. It remains much admired by motoring journalists, and ironically its sales failure at the time makes it a rare and highly collectible model today.

JAGUAR XJ220

ENGINE turbocharged V6

CAPACITY 3498cc

BORE X STROKE 94 x 84mm

COMPRESSION RATIO 8.3:1

POWER 542bhp

VALVE GEAR twin overhead camshafts

FUEL SYSTEM electronic fuel injection

TRANSMISSION 5-speed manual

FRONT SUSPENSION independent double wishbones, coil springs, gas dampers

REAR SUSPENSION independent, double wishbones, coil springs, gas dampers

BRAKES 4-wheel disc

WHEELS alloy

WEIGHT 3025 lbs (1372 kg)

MAXIMUM SPEED 220mph (354kmh)

PRODUCTION 281, 1992-1994

JAGUAR XK8

Jaguar's return to independence in 1984 rescued the brand from the destiny that would befall Rover and MG which remained part of the nationalised British Leyland. New chairman John Egan returned the company to prosperity by a combination of price rises and a focus on established models whose development costs had already been absorbed.

To restore the brand's reputation he brought in Geoff Lawson as Head of Styling. Lawson had previously been Chief Designer at General Motors' British subsidiary Vauxhall. Lawson's efforts deserve much of the credit for the improvement in quality of the company's output.

By the late 1980s Sir John Egan (he had been knighted in 1986 for services to export) was keenly aware that in the face of increased competition from BMW, Mercedes and the Japanese manufacturers, the company might not have the engineering or financial resources to fund the development of future models in the luxury car market. Both GM and Ford were

interested in acquiring the brand potential of Jaguar, and in 1990 Egan accepted Ford's £1.6 billion bid.

America had been a driving force in the company's export and production decisions since 1945 and it was a good match in many ways. Ford's main problem with its new asset was the lack of new product. Jaguar relied more on upgrading existing models, while Ford considered the launch of new ranges, to be essential to high sales and profits.

Jaguar hadn't made a true sports car since the demise of the E-type in 1974, but the XJ-S, a sporting grand tourer launched in 1975, was still doing brisk business in 1990. Nevertheless it was being produced on an ageing 1970s factory line and Ford began the search for a replacement, codenamed X100, as soon as it took possession of Jaguar. By the time that replacement was delivered as the XK8, the XJ-S was 21 years old.

The XK8 coupé was launched, as the E-type had been, at the Geneva Motor Show; the convertible arrived a month later in New York, coinciding with the unveiling of a new exhibit at the city's Museum of Modern Art – that ultimate automotive work of art, the E-type. There were strong echoes of the E-type in Geoff Lawson's shrewd styling of the XK8: the bulge in the bonnet, the glass coverings of the recessed headlamps, the horizontal oval grille. Even the rubbing strip across the doors recalled a similar feature in chrome on the Series 3 E-type, albeit now below the line of the wheel arches. There was, however, a significant absence of brightwork in the XK8's relatively restrained design.

The XK8 was launched in two versions – coupé, and convertible – both as 2+2s, another echo of the Series 3 E-type. Legroom in the rear remained tight, and a large boot was a priority – the soft top was motorised but stowed under a padded cover rather than a metal one which would have eaten into the area of the boot lid. Where the E-type's promotional images showed its back seat occupied by two small children looking over the partition into the boot space, the XK8 publicity material focused on the storage itself, large enough (it showed us) for two sets of golf clubs. This was a sports car, but now the sport was golf and the race was to get onto the fairway as quickly as possible.

The XK8 was significantly larger than the E-type. The interior was all leather, and there were other creature comforts – air conditioning, side airbags, and greatly improved air seals around the windows. Sensors raised the windows a little deeper into the rubber

Left: The Jaguar XK8 had the impressive looks of an Aston Martin DB7 but cost considerably less. Both companies were owned by Ford at the time.

seals when the doors were closed, and lowered them slightly just before they opened. Within the gently sloping rear of the coupé, the characteristic D-shaped rear quarterlight was another reminder of the XK8's antecedents. If the windscreen had sloped more steeply to the scuttle, the similarity would have been all the more striking in the profile of the hardtop.

The dashboard came in maple or walnut veneer, and held half a dozen dials and the heating vents. The switches themselves were in a crowded panel below the dash in a leather bulkhead over the transmission, and some critics felt the sheer number of buttons, rockers and knobs was more Ford than Jaguar in aesthetic. The interior design scheme in general represented a complete break with the XJS, a very different look and feel.

The cosmetics may have been good, but what of the mechanics? Beneath the bonnet, a new Jaguar engine was making its debut. An early beneficiary of the company's new ownership, the AJ-V8 engine was developed by Jaguar and manufactured at Ford's Bridgend plant in South Wales. The 4.0-litre V8 engine was housed in an alloy block, and its smooth running was enhanced by 32 valves for its eight cylinders. It boasted variable cam phasing, and a fast warm-up thanks to its advanced engine management system.

The V8 was seamlessly married to its transmission, at first German manufacturer ZF's 5HP24 five-speed automatic J-gate transmission, electronically controlled. In 2002, when the engine was upgraded to 4.2 litres, Jaguar started to fit ZF's new 6HP26 six-speed box as standard. The XK8 had no manual option.

The vehicle was based on a modified section of the outgoing XJS, which defined the size of the XK8 inside and out. But the new monocoque body was assembled from almost a third fewer panels than its predecessor. The reduced weight, combined with a 25% increase in rigidity, had a very positive impact on its road performance: 0 to 60mph took 6.6 seconds, 0-100 was gained in 16.7, with an effortless progress to over 150mph. Both the coupé and the drop-top were electronically limited to 155mph.

The XK8 was given a suspension system to match this tremendous performance. At the front, double wishbone independent suspension sat on an innovative aluminium subframe derived from aircraft construction technology, with springs set straight onto the body. At the rear, Jaguar's own Computer Active Technology Suspension system (with the knowing acronym CATS)

Above: Between 1996 and 2005 the XK8 convertible outsold the fixed-head coupés by over 2 to 1.

was fully independent with outboard brakes, as used in later X300 saloon versions.

In 1997, the first full year of the XK8's impact on the market, over 14,500 were sold, more than any other sports car in Jaguar's history, out-performing even the XJS on the sales sheets. It was noticeable that many of the XK8's new owners were loyal Jaguar buyers who had previously purchased an XJS, but the car was also attracting drivers new to Jaguar. The success took the company by surprise and delighted Ford, who were investing £200 million in Jaguar's Castle Bromwich factory for the production of the forthcoming S-TYPE saloon. The XK8 was built in Jaguar's home town of Coventry.

Although in the 1980s Sir John Egan had steadily put up the prices of Jaguar cars, placing them firmly in the luxury vehicle market, the XK8 was still remarkably competitive. At around £48,000 for the coupé and £54,000 for the convertible, it was still £15,000 cheaper than its nearest rival, the Mercedes 280SL. The latter's sales were satisfyingly dented by the XK8's success.

The XK8 remained in production for ten years, giving rise along the way to a supercharged version, the XKR. Jaguar made relatively few changes to the XK8's winning formula during its lifetime. Its first significant makeover came in 2001, and that involved little more than new badges and flush-mounted fog lamps outside the cockpit, and restyled seating inside.

The trim was tweaked again in 2002, and Xenon lighting was introduced. The only major change of the car's lifetime also appeared that year – the factory now fitted a 4.2-litre version of the AJ-V8 engine, the AJ34, to both convertible and fixed-head versions of the XK8, accompanied by a new six-speed automatic transmission. Jaguar had made minor improvements to the original V8 in 1998 and 2000, although it had always kept its square bore and stroke, 86 x 86mm. Now the stroke was extended to 90.3mm, resulting in a modest increase in output from 290 to 294bhp.

In 2004 a new choice of seating styles was presented but the most obvious difference from the outside was the enlarged grille and the addition of a full-width air intake below it.

Production of the car was scheduled to end in 2005 to make way for an all-new version of the XK8. But in its final year its supercharged cousin, the XKR, had a string of six championship victories in the SCCA's Trans-Am series. The wins secured Jaguar the manufacturer's title in the North American series for

the fourth time in five years, interrupted only by Ford's triumph in 2002.

To celebrate, Jaguar issued a U.S.-only Victory edition of the coupé and convertible models of both the XK8 and the XKR. It was available in four new metallic colours, and the XK8 models had a new Elmwood interior trim. There were special badges, and chequered flag motifs incorporated into the bonnet badge and the door sills. The XK8 Victory was introduced at the Los Angeles International Auto Show in 2005, for sale in 2006, in a limited edition of only 1050 cars across all four versions, a final Victory lap in the first phase of the XK8 story.

The XK8/XKR Victory proved to be the last new car built at Jaguar's home plant in Browns Lane. Ford's finances were under pressure thanks to a weak dollar and the parent company's investment in Jaguar had yet to see a return. In fact its losses through the British carmaker were running into millions.

In such circumstances sentiment counts for nothing and the line on which Jaguar's workforce had assembled some of the finest cars in automotive history, had to close. The birthplace of the XK120, the Mark 2, the C-, D- and E-types, the XJ-S/XJS and the XK8 was sold. All future production, including that of the forthcoming new XK8, was transferred to the modern Castle Bromwich line in which Ford had sunk so much money. It was a new start, but the end of an era.

When production of the XK8 stopped in 2005 it was by far the most successful sports car in Jaguar's history with combined sales of the coupé and the convertible running to over 66,500. The XKR models accounted for another 23,500. But Ford's financial misfortunes meant that the Browns Lane closure was not going to be enough to secure the company's future.

JAGUAR XK8

ENGINE V8

CAPACITY 3996cc

BORE X STROKE 86 x 86mm

COMPRESSION RATIO 10.75:1

POWER 290bhp

VALVE GEAR twin overhead camshafts

FUEL SYSTEM fuel injection

TRANSMISSION Five-speed automatic

FRONT SUSPENSION independent unequal-length wishbones, anti-dive, coil springs

REAR SUSPENSION independent double wishbones, driveshafts used as upper links, coil springs, anti-squat (CATS option)

BRAKES 4-wheel ventilated discs

WHEELS alloy

WEIGHT 3560 lbs (1615 kg)

MAXIMUM SPEED 155mph (249kmh)

PRODUCTION 66,518, 1996-2005

K8 XKO

JAGUAR XKR

The XK8 was an immediate success when it was launched in 1996. It looked remarkably similar to an Aston Martin DB7 but was around £40,000 cheaper. In its first full year of production, 1997, it broke all Jaguar's own sales records. In 1998 the XKR appeared, outwardly almost identical to its more sedate cousin but inwardly altogether more aggressive. The new car would now equate to the Aston Martin on performance as well as looks.

The addition of an Eaton supercharger to the standard AJ-V8 engine pushed the already powerful V8's output up from 290bhp to 375bhp. This was more power even than the V12 first fitted to the Series 3 E-type, and would propel the car from 0 to 60mph in 5.2 seconds.

To feed this hungry machine the front grille, which on the XK8 had a horizontal splitter bar, was one full-width mesh panel on the XKR. In addition, louvre panels were set into the bonnet either side of the central bump. At the other end, the XKR exhaust was of a wider bore than the XK8's.

Like the XK8, the supercharged car was limited to 155mph. To improve handling, Jaguar fitted Servotronic steering and better disc brakes as standard. The XKR had the same sophisticated Jaguar CATS suspension system, first seen in the X308 saloon.

Below: Apart from a very subtle rear spoiler, the XKR had a wire mesh front grille. The major difference was under the bonnet.

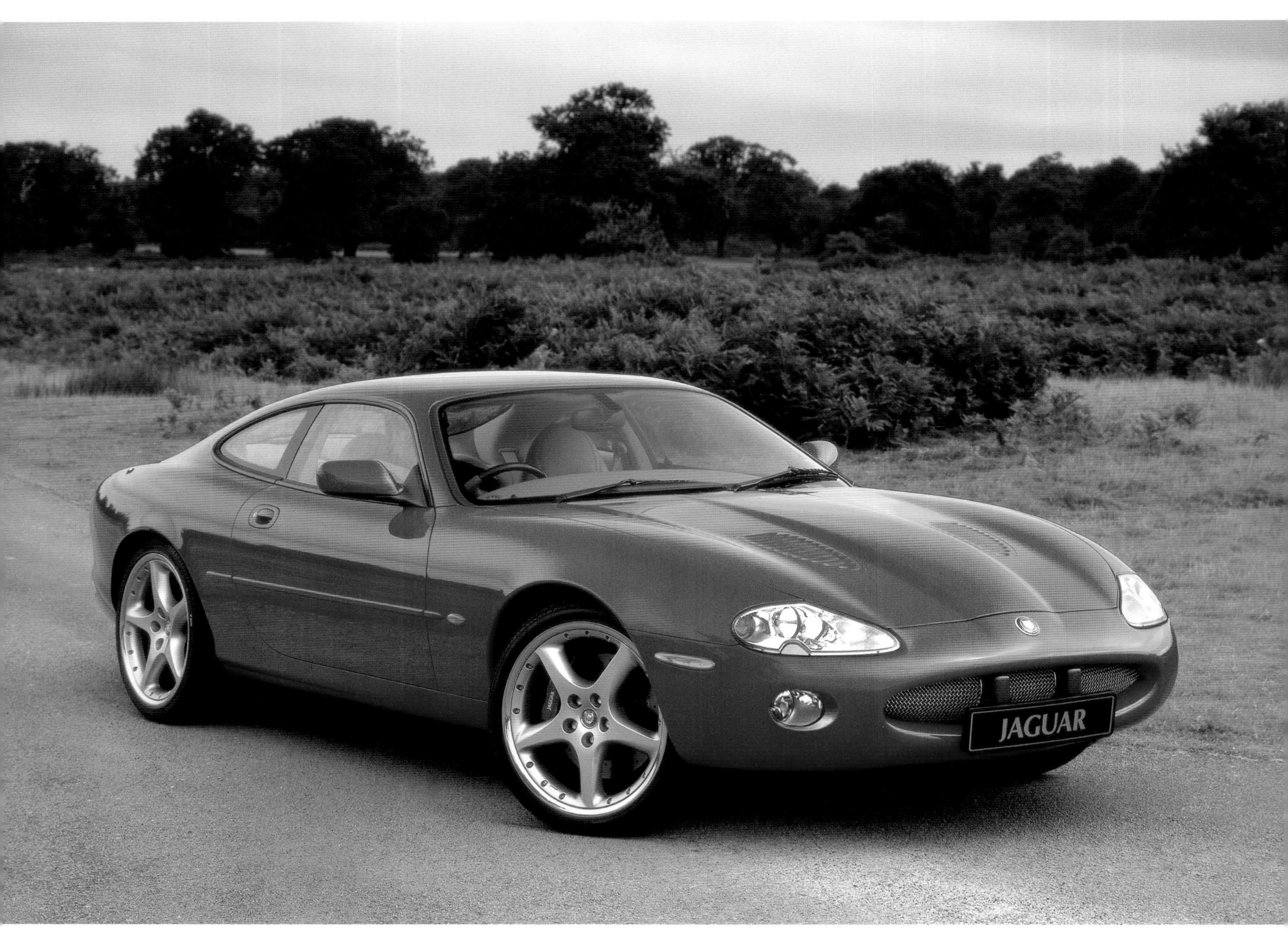

There were minor modifications to the body of the car too, all designed to improve stability and handling. To reduce drift on corners, the XKR was the first Jaguar to be fitted with wider wheels at the back than at the front, and to accommodate them, the rear arches were given rubberised extensions. The rear also received a small aerodynamic spoiler at the back of the boot lid. New badges on the nose and the boot bore the word 'supercharged'.

Transmission was initially different in the XKR – the ZF five-speed automatic system in the XK8 was replaced by a Mercedes W5A580 package, until in 2002 an upgraded ZF six-speed transmission became standard for both cars. In the same round of improvements, the V8 was re-bored to 4.2 litres, giving the supercharged model an output of 400bhp.

Throughout its life the XKR got the same changes in interior trim as the XK8, and in addition an almost annual redesign of its alloy wheels depending on what option was chosen – most years anything from 17" to 20" wheels were available. Between them, in their lifetimes, the XK8 and XKR were supplied with at least 16 different styles of wheel. In 2004 the XKR got one final round of modifications when the rear spoiler was enlarged and the twin exhausts of the earlier version became quadruple pipes.

Apart from these regular upgrades, the XKR was the basis for a number of special editions, generally celebrating a Jaguar milestone, anniversary or event. The first event, and probably the one which most gladdened the hearts of Jaguar enthusiasts around the world, was the company's return to topline motorsport.

In 1999 Ford decided that the Jaguar brand would be the company's presence in Formula One from the 2000 season onwards. The parent company bought world champion Jackie Stewart's racing team Stewart Grand Prix and renamed it Jaguar Racing. To celebrate their entry into Formula One, Jaguar revealed the XKR Silverstone. Built in a very limited edition of just 50 coupés and 50 convertibles, the XKR Silverstone was only available in high gloss platinum silver – contrasted by the convertible's all-black hood. Its 20" BBS Detroit alloy wheels, an inch wider at the rear than at the front, gave it an imposing presence. Combined with Brembo disc brakes the increased rear track gave the Silverstone the best handling in its class.

The new owners were invited to attend Jaguar's practice sessions at Silverstone on April 13th 2000. Jaguar raced in Formula One for the next five seasons, but with very little success, and in the same round of cost-cutting measures which saw Ford close the

Below: The XK8/XKR was the first Jaguar to use the new AJ-V8 engine, continuing the long tradition of Jaguar trialling important engines in their sports cars.

Browns Lane plant it sold Jaguar Racing on to soft drinks magnate Dietrich Mateschitz, who rebranded it Red Bull Racing.

In 2001 Jaguar brought out another special version, the XKR100, to mark the centenary of the birth of its founder Sir William Lyons. This version was available in a run of 500, equally split between the convertible and the coupé, mostly for the North American market – only 80 of each remained in Britain.

The XKR400, which appeared in 2003, was distinguished mainly by its choice of three colours – platinum silver, slate grey, and midnight black. It marked an upgrade to the XKR's engine and transmission, and was only available to certain UK dealers, who could each order a maximum of five cars from the edition. Only 100 in total were produced. While still more expensive than the standard XKR models, the XKR400 was cheaper than previous limited editions, and was quickly snapped up.

Lest North America should feel left out by the UK-only XKR400, 2003 also saw a release limited to the U.S., the XKR Portfolio. It was available in two bright metallic finishes, Jupiter Red or Coronado Blue, each boasting matching interior leather (including a leather steering wheel) with charcoal detailing on Recaro front seats. Another limited edition of 100, known as the Carbon Fibre model, had the usual veneered dash fascia replaced with a carbon fibre one.

A five-car coupé-only edition carried the signature of Stirling Moss. Moss had begun his racing career in a Jaguar XK120 at Dundrod 50 years earlier, and the five XKR Stirling Moss cars, one for each decade, honoured his contribution to Jaguar racing history. They came in silver with Jaguar's twin Racing Green stripes running back from the nose to the Borla exhaust system, Super-Sports suspension and tinted windows. At 466bhp it was the most powerful factory Jaguar since the XJ220.

Finally, to celebrate the XKR's success in the SCCA's Trans-Am series, one last special edition, the Victory, was produced in 2005 across both the XK8 and XKR ranges. As well as the full production range of colours, the Victory was available in four new metallic shades: Satin Silver, Black Copper, Bay Blue and Frost Blue.

In 2005 Ford closed Browns Lane. The last cars were XKRs, and the final 30 of them were designated the XKR Stratstone 4.2S edition. They were effectively the same as the Victory edition. But each had a uniquely numbered and polished door footplate with a chequered flag motif on the plate and on the special white Jaguar badge. And there was no denying their collectability – no further cars would ever be assembled in Jaguar's historic home, they were the last of the line.

JAGUAR XKR

ENGINE supercharged V8

CAPACITY 3996cc

BORE X STROKE 86 x 86mm

COMPRESSION RATIO 9:1

POWER 375bhp

VALVE GEAR twin overhead camshafts

FUEL SYSTEM fuel injection

TRANSMISSION 5-speed automatic

FRONT SUSPENSION independent unequal-length wishbones, anti-dive, coil springs

REAR SUSPENSION independent double wishbones, driveshafts used as upper links, coil springs, anti-squat

BRAKES 4-wheel ventilated disc

WHEELS alloy

WEIGHT 3616 lbs (1640 kg)

MAXIMUM SPEED 155mph (249kmh)

PRODUCTION 23,856, 1998-2005

JAGUAR

JAGUAR S-TYPE

During Jaguar's brief period of independence in the 1980s, managing director John Egan had pushed the company firmly towards the luxury car sector. But once the company passed into Ford's hands in 1990 the new owner had a very different strategy. The luxury sector was not large enough by itself to sustain the marque, and Ford was keen for Jaguar to increase its sales in the larger and highly competitive market of the mid-range executive vehicle.

In this sector the BMW 5-Series was king, and Jaguar had not competed in it since production of the beloved Mark 2 had ended in the late 1960s. To give an idea of the scale of the potential new market, Ford hoped that Jaguar's new entry would double Jaguar's sales figures. This was a prize worth competing for.

Jaguar's award-winning chief designer Geoff Lawson was put in charge of styling the project, and Jaguar were given complete budgetary control over it. But their hands were to an extent tied by the political expediency of sharing components with other parts of the Ford empire. To save on costs the car was developed alongside another new model, Ford USA's Lincoln LS. The two projects shared a floorpan, suspension and other mechanical aspects including in some versions a Ford engine.

Lawson took the Mark 2 as his starting point. The S-TYPE name (now in upper case) harked back to the 1963 car of the same name which was itself derived from the Mark 2. But the design process was not a smooth one and four prototypes were rejected before the fifth at last became the S-TYPE that was introduced to the public at the Birmingham Motor Show in 1998.

It was design-by-committee, or perhaps by management. Lawson's first effort, a rounder, scaled-down version of the XJ, was turned down for not looking different enough for the new market – it was, apparently, too much like an old luxury Jaguar. His second, a saloon version of the existing XK range, was blocked for not being 'Jaguar' enough. A third approach was 'very Jaguar', reclaiming the old vertical slatted grille and aping the Mark 2 profile with a droopline that highlighted a large bootspace.

Lawson's fourth attempt was, at last, half-right; the front half would eventually be retained in the final design. But Lawson, who drove a red 1969 Chevrolet Corvette to work, had stretched the form and made the S-TYPE too 'American'. Above all, Ford wanted their executive Jaguar to look 'British'. Finally, S-TYPE number five ticked enough boxes to be given the go-ahead. It had the front of the fourth version, the Mark 2 profile of the third, and yet another new rear end.

There were traces of Jaguar styling throughout – the rippled bonnet leading to the headlamps echoed the wings and headlamp fairings on the Mark 2 and the original S-type. The slatted grille, although much shorter in height now, was definitely a Jaguar grille, even if without a central spar it was closer to the Mark 1 than the Mark 2. Chrome corner flashes mimicked the split front bumper of the original S-type, but without over-riders. There was a vestigial D-bend about the rear quarterlight, although it now met the bottom edge of the side windows almost at a right-angle, not a smooth returning curve. But the rear end, whatever the Detroit management thought, was described by some as more Daewoo than Jaguar.

The interior also attracted some criticism, despite being finished to Jaguar's usual high standards in leather and walnut veneer. A hump in the dashboard to accommodate the display behind the steering wheel broke with tradition; and the mass of plastic controls

Below: The S-TYPE that was unveiled at the 1998 Birmingham Motor Show was the fifth version of the car that Geoff Lawson's design team had produced.

housed in a large U-shaped panel below the dash and above the gearshift also offended some. Others found the steering wheel itself, finished in wood and leather, too showy.

All these jarring details were eventually addressed in a major restyling exercise by Ian Callum in 2002 which coincided with a series of technical upgrades. Electronic pedal adjustment was introduced in all models at that time, and electronic seat adjustment in the higher-end models.

The process must have been very frustrating for Lawson, whose work with Jaguar since 1984 had won many awards. He died a year after the launch of the S-TYPE, aged just 54. He suffered a stroke after making a design presentation at the company's Whitley Engineering Centre. A workshop at Whitley is now named the Geoff Lawson Studio in his honour and Jaguar have also endowed a Geoff Lawson Scholarship at London's Royal College of Art in his memory. The sad irony is that Jaguar's previous inspirational design chief, Malcolm Sayer, also died at 54.

Motoring style correspondents were contemptuous of the final product, accusing it of being retro by numbers, a pastiche of a classic car rather than a modern successor to one. The Rover 75, launched at the same show as the S-TYPE, drew much more favourable reviews. But, to Ford's credit, the car-buying public loved it. Not only did it compete successfully with the executive models of BMW and Mercedes; it gave Jaguar owners who wanted to downsize somewhere to go without being disloyal to their brand of choice. The S-TYPE did exactly what Ford had intended, open up new markets for the company without affecting Jaguar's sales in its traditional luxury and sports car strongholds.

In its lifetime the S-TYPE appeared in 22 different versions, supplied with one of five different high-performing engines. At launch it came with either Jaguar's old 4.0-litre AJ-V8 or a new 3.0-litre V6. The latter used the block and crank of a Ford Mondeo engine, but was otherwise beautifully re-engineered by Jaguar technicians to deliver 240bhp at 6,800 rpm.

The V6 was made at Ford USA's Cleveland plant, the V8 in Ford's factory in Bridgend, South Wales. Thanks to Ford's massive investment, the S-TYPE was the first Jaguar to be fully built on Jaguar's Castle Bromwich production line, which had previously only been used to assemble and paint bodywork.

The V6 models came with two Ford-designed transmission options: either a new five-speed manual box built by Getrag in Germany, or a four-speed automatic from Ford USA.

The V8 had proved itself in both the XK8 and the XJ series, and continued in the S-TYPE to be married to ZF's 5HP24 five-speed automatic J-gate transmission. As with the XK8, there was no manual option for the 4.0-litre S-TYPE. From 2002 the AJ-V8 was upgraded to 4.2 litres, in a standard and a

JAGUAR S-TYPE

ENGINE V6 or V8

CAPACITY 3996cc

BORE X STROKE 86 x 86mm

COMPRESSION RATIO 10.75:1

POWER 281bhp

VALVE GEAR twin overhead camshafts

FUEL SYSTEM fuel injection

TRANSMISSION Five-speed automatic

FRONT SUSPENSION independent double wishbones with alloy control arms, coil springs, anti-dive

REAR SUSPENSION independent double wishbones with alloy control arms, coil springs, anti-dive (CATS option)

BRAKES 4-wheel ventilated disc

WHEELS alloy

WEIGHT 3805 lbs (1726 kg)

MAXIMUM SPEED 150mph (241kmh)

PRODUCTION 291,386, 1999-2007

Above left: The supercharged version of the S-TYPE, the S-TYPE R, was equipped with an Eaton M112 supercharger. In 2004 an economical 2.7-litre diesel engine that had been jointly developed by PSA (Peugeot/Citroën) and Jaguar/Land Rover was added to the range.

new supercharged form. Aspiration of the latter was assisted by a new mesh front grille, and both forms now used ZF's six-speed 6HP26 transmission. At the same time the 3.0-litre V6 was joined by a 2.5-litre version, although this was dropped again three years later, soon after the introduction for the first time of a 2.7-litre diesel engine for the range.

The S-TYPE's great contribution was the development of a completely new suspension system, the first since that of the XJ nearly 40 years earlier. A 2002 revamp also saw a move to laser assembly of the body panels. Besides that year's major revision of the interior, the S-TYPE saw little change beyond some wing mirror and badging detail until 2004. Then a number of small but improving adjustments to the outer appearance were made: re-shaped bumpers front and back, and a higher tail giving a flatter boot lid. The doors sills were more clearly defined and Jaguar dispensed with the side rubbing strips. The overall effect was to simplify the lines of a car whose original design had been a compromise, a composite of elements from several design stages.

In the final couple of years of the S-TYPE, Jaguar followed the pattern with the XK8 of releasing a series of special editions to maintain sales levels. In 2006 the standard diesel, 3.0- and 4.2-litre models were upgraded with many extra features and rebranded as the S-TYPE Spirit. In this form it served out its days until discontinued in 2007 when the XF was ready to launch. The S-TYPE more than fulfilled the brief from Ford. In its 22 incarnations Jaguar sold over 290,000 of them to its new executive customer base.

Below: The S-TYPE received a facelift in 2004 courtesy of Geoff Lawson's successor, former Aston Martin and TWR designer, Ian Callum, who took over as design director in 1999.

JAGUAR XK/XKR

Although usually described as the updated XK8, it was made clear from the XK's launch that it had very little in common with its predecessor. Its build and style were based on that of a concept Jaguar, the ALC (Advanced Lightweight Coupé), which was shown at the North American International Auto Show in Detroit in January 2005.

The ALC was designed by Ian Callum, and continued a long-standing Jaguar tradition of drawing inspiration from the weight-saving, rigidity-enhancing construction techniques developed in the aerospace industry. It was built entirely of aluminium, and a flexible approach to assembly meant that joints were bolted, glued and riveted together as appropriate; in fact there was only one conventional weld in the whole body. The result in the new XK was a car 50% more rigid than the XK8 despite being 20% lighter.

The convertible was launched at Detroit the year after the ALC had been shown and a few months after the XK coupé's debut at Frankfurt in 2005, but Jaguar made it clear that the convertible had been designed before the coupé to guarantee its rigidity independent of a fixed-head version. Both models were 2+2s, very much in the E-type mould with relatively little comfort in the rear. The XK had a longer wheelbase but was shorter beyond the axles at both ends than the XK8, and storage was the priority at the rear, not seating. There was room enough for those theoretical two sets of golf clubs in the modest boot – a boot which in both formats now had a mini-spoiler with built-in brake light.

The shorter styling also recalled the E-type; there was a smoother, rounder curve to the front of the coupé's cockpit, which had the E-type's characteristic

Right: The Advanced Lightweight Coupé as it appeared at the 2005 Geneva Motor Show.

hatchback rather than the XK8's boot lid. Callum favoured a retro soft-top for the convertible. His triple-layered fabric roof could now be raised or retracted electro-hydraulically in just 18 seconds, and stowed itself fully under an aluminium panel.

There were some innovative safety features. Rollbars automatically rose over the rear of the convertible's accommodation at the first sign of a potential accident. There were twin airbags, for head and thorax, and motion sensors would push the headrests forward to support the neck in an impending impact, reducing the risk of whiplash. Another device, unique at the time, was designed to protect not the occupants but any pedestrian struck by the car. In such an event, pyrotechnics raised the thin aluminium bonnet so that the dent in it, caused by the weight of a body, would not bring the body in contact with the hard mechanics

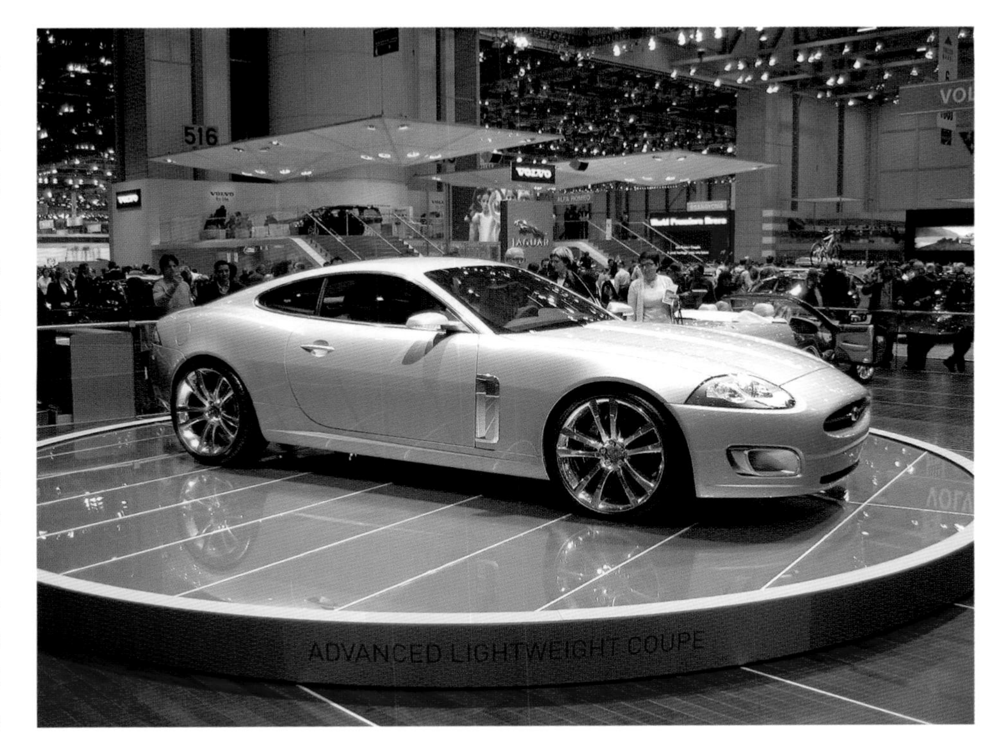

beneath it. This thoughtful feature was prompted in part by the aluminium skin used on the XK, which was lighter and more pliable than ever.

The lightweight body was supported on intelligent suspension which gave the XK Jaguar's smoothest ride yet. New unequal double wishbones front and back were linked to a control system denoted as Adaptive Dynamics, which ran electronically adjustable shock absorbers to regulate vertical movement as well as pitch and roll. The driver could select a more dynamic setting for a racier feel. Another box of tricks called Dynamic Stability Control collected information from traction sensors and anti-lock brakes, with which it monitored and managed wheelspin, reining in any under or oversteer.

Not surprisingly, other electronic luxuries came as standard: satellite navigation, climate control and telephone, all controlled from a seven-inch screen in the dash; cruise control, parking guides and rain-sensing windscreen wipers; keyless entry and ignition, and automatic locking on exit were all included. A further £6000 of extras were available in items like tyre pressure monitoring and a heated steering wheel. Active lighting, another optional extra, helped to illuminate the road ahead by following the movement of the steering wheel.

At launch, both models of the XK were supplied with the trusty AJ-V8, which had made its debut as a 4.0-litre machine in the XK8 in 1998. Designed by Jaguar and built at Ford's South Wales factory in Bridgend, the V8 was now produced as a 4.2-litre unit.

Above and Left: The Jaguar XK 4.2. Viewed from above the rear window was highly reminiscent of the E-type Jaguar.

If it were not limited to 155mph, it could certainly have done more. Because of the reduced weight of the new XK, even the naturally aspirated version fell only a quarter of a second short of its supercharged predecessor over a quarter-mile sprint; and for a high performance vehicle its fuel economy was not so bad at 25mpg.

In October 2006 the new supercharged XKR made its debut. It too was limited to 155mph, although some thought that, unrestrained, it was capable of at least 180mph. Its 0 to 60mph time of under five seconds suggested as much. Eaton boosters helped to push the AJ-V8's output up to 420bhp. They augmented an already expanded natural air supply. The XK range incorporated the classic low, wide, oval grille of the E-type, and introduced a future trademark Jaguar feature, a vertical louvre panel on the front wings just forward of each door. The XKR had additional vents on the bonnet, and its emissions were from four tail pipes compared to the XK's twins. Its bumpers were restyled and the bold horizontal chrome bar which ran across the XK grille was eliminated on the XKR. The uninterrupted mesh mouth suggested an unsmiling, more aggressive approach to aspiration but left the 'growler' Jaguar badge looking a little lost in its centre.

In both standard and supercharged transmissions, the V8 continued to be linked to the six-speed ZF 6HP26 gearbox, as it had been since it was upgraded to 4.2 litres in the XK8. The box had been the world's first six-speed automatic when it was introduced, and its smooth transitions still made it the obvious choice for a car whose ride was so finely and electronically tuned. The XK and the XKR both featured paddle shifts on the steering wheel, a degree of manual control for drivers who liked to take a more active part in driving such a powerful machine.

The XKR was the starting point for several limited editions. Jaguar's policy in the past had been to roll out special versions towards the end of a car's life to provide a boost to falling sales. But the first such event for the XKR arrived only a year after the supercharged car's launch. It was called the Portfolio and, perhaps as a nod to the final production of the E-type, it was available only as a glossy black coupé (except in the UK and Switzerland where it was also offered in silver).

The Portfolio showed changes to details of the interior trim. Outside, it had 20" wheels and bigger, better brakes; and the louvre vents on the bonnet and wings were restyled. An XKR Portfolio Convertible joined its hardtop brother a year later, and both cars

Below: The twin tailpipes make it quite clear this is an XKR.

DII ANE
H-A-FOX

remained in production for a couple of years. They were briefly joined by the XKR-S, a European limited edition of only 200 cars.

Launched at the 2008 Geneva Motor Show (scene of the E-type's introduction to the world all those years ago), an XKR coupé with adjustments in engine management, suspension and aerodynamics made it the fastest production model since the XJ220, capable of speeds up to 174 mph.

The most important event in the XK's lifetime was the arrival in 2009 of the next generation of the AJ-V8. Now a five-litre unit, its output was up by nearly 100bhp, to 510bhp in the XKR and 385bhp in the standard XK. Jaguar launched revised versions of both the XK and the XKR for 2010, carrying the five-litre engine alongside improved steering and braking systems, and with styling changes inside and out. The gearstick was replaced by a circular gear selector that rose up, out of the central console when the engine was started.

At the 2011 Geneva show Jaguar launched a more generally available version XKR-S coupé. Besides improvements along the lines of the earlier limited editions, the new engine gave the new model 542bhp and a top speed of 186mph, making it easily the most powerful production car in the Jaguar range. Elements of its more aggressive exterior styling were carried over into facelifts for the other XK models over the next year, and to complete the range an XKR-S Convertible was unveiled in 2012.

Numerous special editions of the XKR were released in the model's final years. To mark the company's 75th

Above and Right: Two different versions of the XK cabin and two different transmission options. The wood-veneer dash at right has the older J-Gate shifter.

JAGUAR XK/XKR

ENGINE V8

CAPACITY 4196cc

BORE X STROKE 86 x 90mm

COMPRESSION RATIO 11:1

POWER 294bhp

VALVE GEAR twin overhead camshafts

FUEL SYSTEM fuel injection

TRANSMISSION six-speed automatic

FRONT SUSPENSION independent unequal-length wishbones, coil springs

REAR SUSPENSION independent unequal-length wishbones, coil springs

BRAKES 4-wheel ventilated disc

WHEELS alloy

WEIGHT 3516 lbs (1595 kg)

MAXIMUM SPEED 155mph (249 kmh)

PRODUCTION 51,669, 2007-2014

anniversary in 2010 the XKR75 (in the UK, launched at Goodwood) and the XKR175 (in North America, revealed at Pebble Beach) were created in limited numbers of 75 and 175 respectively. Increased top speeds, better aerodynamics and larger rear spoilers were their main features, with different details of trim for the different markets.

As production came to an end in 2014, the final 50 cars were of course given special treatment, based on two earlier editions: 2013's XKR-S GT (produced in a limited run of 40), and 2014's XK Dynamic R. For the final edition, 25 coupés and 25 convertibles, all for the North American market, were given luxurious interiors and additional exterior features – a rear diffuser, wider sills, and extra louvres in the bonnet.

The most exclusive version of the XKR was an earlier one, the 2011 Poltrona Frau edition, based on the personal choices of the car's designer Ian Callum. It sported a red and black exterior, including gloss black wheels; and inside, charcoal leather inside with cranberry red stitching and piano-black veneer. Equipped to produce 510bhp, this ferociously powerful combination is now owned by (Callum aside) just 18 lucky drivers.

Left and Below: The Jaguar XKR convertible. In 2012 it was joined by the aggressively-styled XKR-S Convertible with a prominent rear spoiler.

JAGUAR XF

The successor to the S-TYPE executive saloon, the Jaguar XF, was launched at the Frankfurt Motor Show in 2007. It appeared after the C-XF concept had been revealed in January 2007 at Detroit. Jaguar's future rested on the success of this car. In 2007, sales for the company stood at only a quarter of their 2002 levels, down from 61,204 cars to only 15,647. By the time the XF was launched, Jaguar was widely perceived by industry experts to be in its death throes. After a promising start, the S-TYPE was failing to hold its own against its competitors in the executive luxury market, Mercedes and BMW. The gamble of the XF was to combine the attitude of a sports car with the practicality of a family saloon and to preserve the best of both styles of vehicle.

The S-TYPE had already made some inroads into the market before its sales declined. With the launch of the XF, Jaguar restated its intent, returning to the production of premium cars for niche markets. It was immediately perceived as a viable alternative to the Mercedes E-Class and the BMW 5-Series, and sales in the first year of XF production outperformed predictions. With figures of 30,000 in its first twelve months, it outsold its S-TYPE predecessor by two to one.

The XF marked a new phase for Jaguar in two senses. In design terms it took the heritage of the marque, with its sense of style and refinement, into the 21st century. At the hands of designer Ian Callum the new car demonstrated an evolving, modern design philosophy which embraced high-tech specifications in engine

Below: The Jaguar C-XF concept revealed at the North American International Auto Show at Detroit in January 2007.

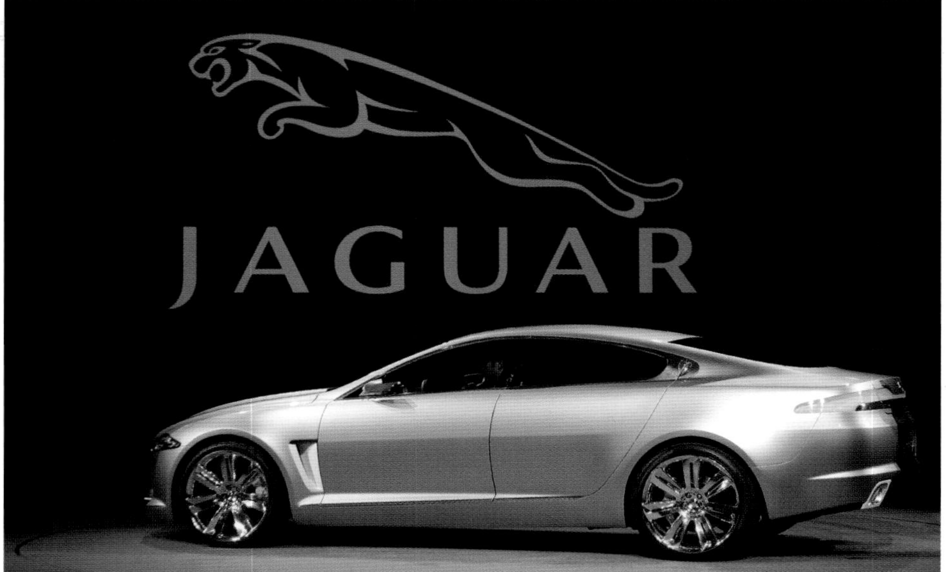

JAGUAR XF

ENGINE V8

CAPACITY 4196cc

BORE X STROKE 86 x 90mm

COMPRESSION RATIO 11:1

POWER 298bhp

VALVE GEAR twin overhead camshafts

FUEL SYSTEM fuel injection

TRANSMISSION 6-speed automatic

FRONT SUSPENSION independent unequal-length wishbones, coil springs

REAR SUSPENSION independent unequal-length wishbones, coil springs

BRAKES 4-wheel ventilated disc

WHEELS alloy

WEIGHT 3856 lbs (1749 kg)

MAXIMUM SPEED 155mph (249kmh)

PRODUCTION 2008-

performance, aerodynamics, safety and comfort.

The XF also represented a new phase in the ownership of the marque. When Callum designed it at Jaguar's Whitley HQ, it was still owned by Ford. By the time the first cars were rolling off the production line in March 2008, the company was owned by TATA, India's largest car manufacturer.

Jaguar had never made a profit under Ford's ownership, and as the global financial crisis loomed Ford needed to slim down its operations. TATA, the world's fourth largest manufacturer of trucks and second largest of buses, was an unlikely destination for the luxury marque. While negotiating with Ford to acquire Jaguar, TATA was also announcing the launch of the Nano, the world's cheapest car.

But despite the contrast in the companies' vehicles, TATA (established in 1945) was the preferred buyer, with its respectably long history and successful volume sales. The Indian company was seen as a safer pair of hands for the venerable marque than the venture capital deals also on the table. For their part the new owners must have been delighted with the XF's early success in its intended market: it won *What Car* Best Executive Car four years running from 2008-2011, and was named *Auto Express* Car of the Decade in 2011.

The XF was promoted as a 'saloon within a coupé'. Its coupé style, with its deep mesh front grille, harked back to the original 1968 XJ6. But that was the only nod to the past. Ian Callum was determined that the XF would embody a new, forward-looking aesthetic, in contrast to the retro styling of the S-TYPE which, ironically, had dated quite quickly. In the XF, it was the technological rather than aesthetic heritage of Jaguar that brought dividends.

Ian Callum made use of CFD (Computational Fluid Dynamics) in the design, which allowed every element of the external profile to be optimised for aerodynamic efficiency before a prototype entered the wind tunnel. As a result, with a drag coefficient of 0.29, and no front-to-rear lift at all, the XF became (against tough historic competition) the most aerodynamically efficient Jaguar yet produced.

The XF did not cut all ties with its predecessors, though. It used the S-TYPE's aluminium chassis, but with structural changes that delivered improvements in safety, stiffness and space efficiency. The carbon steel shell, although heavier than the alloy shell of the XK, gained in rigidity what it lost in lightness.

The body was lengthened slightly in order to comply with the latest crash safety requirements. Additional headroom was created by a higher body, without sacrificing the 'saloon within a coupé' proportions. The engine line-up was also very similar to the S-TYPE, whilst the suspension and mountings were similar to those of the XK.

Suspension and drivetrain also benefited from

Above: The XF took the remarkable accolade as *What Car*'s Best Executive Car from 2008 to 2011.

technological innovations first launched in the S-TYPE. The lightweight, anti-dive wishbones of unequal length at the front of the car, and the multi-links at the rear delivered excellent ride experience and bump absorption even at high speed. Jaguar's Adaptive Dynamics system of sensors, first seen in the XK, adjusted the damping in line with driving style. Active Differential Control regulated the rear differential electronically.

Transmission was originally through a six-speed automatic gearbox, an eight-speed ZF box arrived in 2011. Power came from the AJ-V8, initially in its second-generation, 4.2-litre spec. When the 5.0-litre third generation version was rolled out in 2009, the XF was upgraded along with its stable companions the XKs. An All-Wheel Drive option was introduced in some markets, with the 3.0-litre V6 engine.

Above: A styling cue from the XK, the engine side vent, or 'Power Vent', can be enhanced in higher specification models, where it can be supplied in carbon fibre.

In 2009 Jaguar launched a supercharged XFR with an output of 503bhp. The XFR came with a rear spoiler and 20" alloy wheels. Later that year, on the Bonneville Flats, Paul Gentilozzi drove a modified XFR at 225.675mph, smashing the Jaguar record of 217.1 mph held by the XJ220 since 1992. Any suggestion that Jaguar's sporting potential was all in the past could now be forgotten.

The XF, conceived for a global market, was launched with a variety of trims depending on the destination country, but none of them were of cloth. Jaguar positioned the car firmly as a premium model, and even entry-level versions were fully trimmed in leather and veneer. More recently a choice of veneers has been accompanied by options for carbon fibre, aluminium and piano-black lacquer.

In 2011 the whole XF range was given a modest facelift, and expanded to include a 2.2-litre diesel model, aimed squarely at the company fleet market. In an unusual collaboration, the diesel engines for the XF were developed in partnership with Peugeot-Citroen.

The following year Jaguar had a second crack at a difficult nut, the estate car. Jaguar's first (official) attempt at this market was the X-TYPE Estate launched in 2004, and also designed by Ian Callum. In the 1970s and 1980s private firms had taken the XJ-S and coachbuilt their own 'sporting brake' cars with varying degrees of success. The Jaguar X-TYPE had been very much a Ford product, manufactured at their Halewood assembly plant and based on the Ford Mondeo platform known as the Ford CD132. As a branding exercise any link to the Mondeo robbed Jaguar of its automotive cachet. The car, known within Jaguar as the X400, was in production from 2001 to 2009. The X-TYPE Estate was widely perceived as being the most attractive model in the X-TYPE range, and was one of the last to be discontinued at the end of the X-TYPE's run. It had sold moderately but had failed to oust the equivalent models of its main competitors BMW and Mercedes.

Top: Mark Cavendish and Juan Antonio Flecha of the Sky team ride alongside their new support car for the 2012 season, the Jaguar XF Sportbrake

Above: The Jaguar XF Sportbrake launched at the 2012 Geneva Motor Show.

The new XF Sportbrake was available with a choice of diesel engines, either 2.2 or 3.0 litres. Its load capacity swallowed that of its competitors at Mercedes and BMW: folding down the rear seats trebled the available space from 550 to 1,675 litres. It was widely hailed as a car that delivered practicality in terms of payload volume but did not compromise on speed and handling or on aesthetics. The Active Electronic Differential

Above: The Sportbrake posed outside Eastnor Castle in Herefordshire.

Left: The 2015 XFR-Sport in Rhodium Silver.

and Jaguar's own Dynamic Control Stability Systems from the XFR-S saloon meant that, despite being an estate, it handled like a sport saloon, and thanks to Ian Callum its curves avoided any suggestions of boxiness. It remains in production and has outsold its X-TYPE forebear.

The second generation XF was launched in 2015, with a stunt worthy of James Bond, crossing the water at London's Canary Wharf on a high wire. The performance was designed to show off the car's new lightweight construction, now 75% aluminium, which reduced the weight by 190kg, with associated fuel efficiency and reduced CO_2 emissions.

The body was aerodynamically revised, with new bumpers and now horizontal side vents in the front wings. It was 7mm shorter than the first generation XF, but a wheelbase longer by 51mm delivered extra legroom in the back; there was more headroom too. The new architecture also offered 50:50 balance across the axles, improving road-handling.

A raft of intelligent software features to manage handling and fuel efficiency were included as standard. On-board computers reacted intelligently to road conditions, creating predictive as well as reactive adjustments to maintain traction. With JaguarDrive drivers could personalise steering, throttle response and gearbox shift points.

Under the bonnet, Jaguar simplified choice (at the start at least) by dispensing with the old V8. Instead, the 3.0-litre V6 was supplied in supercharged form, and a new 2.0-litre version was a budget option. The XF was now available with Jaguar's new flagship Ingenium 2.0-litre four-cylinder diesel engine. The new take on the XF was well received, and the improved comfort and the benefits to handling of all that electronic wizardry were widely welcomed. The XF was proving to be a thoroughly 21st century car.

XF
AJ09 YJG

JAGUAR XJ

Ever since the launch of the XJ series in 1968, it has been the premier saloon car in the Jaguar range. Though a wholly different motor car, the styling of the XJ8, on sale till 2009, still bore a remarkable resemblance to the 'XJ40' model of 1986. The company's latest large luxury vehicle would break the mould and the familial link. The new XJ was announced in 2009 as the successor to the XJ8. Although it carried forward the XJ tag first used in the 1960s, its styling by Jaguar's chief designer Ian Callum continued to look forward not back.

Its choice of supercharged V6 and V8 engines maintained Jaguar's interest in high performance, but the overriding impression from the XJ was one of elegant luxury. The car was designed, as one reviewer put it, not to quicken your pulse but to relax it. It was still very British in spirit, not as showy as its Italian rivals or as traditional as the German opposition.

The appearance of the new car made little or no reference to any of its XJ predecessors. Instead, it took on board some aspects of the company's medium luxury car the XF, by which Callum had already declared his intent to break with the past. It took the XF's front, for example, and exaggerated it with a bigger grille and more pronounced shaping of the wings which emphasised the line of the bonnet: an altogether more aggressive look. The already streamlined cockpit of the XF saloon became almost a coupé on the new XJ, and now incorporated a large glass roof as standard. This allowed the roofline to be lower than ever, while the extra light which the roof now let in countered any corresponding sense of claustrophobia.

The rear end was radically different from earlier models. Gone was the roundness forever associated with the E-type and the Mark 2. Much simpler lines now defined a vertical tail, with the number plate now held below the bumper and only a leaping cat Jaguar motif interrupted the fall of the boot lid above it. The longer slope of the back window overhung the bootspace, and this arrangement maximised the size of the boot opening, which was emphasised by the elongated light units either side of it.

The lower roof and the aluminium monocoque construction of the body (reported to be around 50% from recycled sources) contributed to a drag coefficient of 0.29, the same as the XF. Those aircraft construction

techniques made the XJ up to 150kg lighter than its rivals in the executive car market. The BMW 550i, for example, was 115kg heavier.

At launch the XJ was offered with a choice of three drivetrains – 3.0-litre diesel, 5.0-litre normally aspirated petrol, and 503bhp supercharged 5.0-litre petrol in the Supersport. All were linked to the ZF six-speed automatic gearbox with JaguarDrive's paddle controls on the steering wheel for manual shift.

Driving the XJ was further refined by a raft of intelligent systems to control brakes, steering and suspension; 21st century technology which Jaguar had first started to embrace in the XK sports range. After decades of seeing itself as virtually a heritage brand, Jaguar was repositioning itself as the contemporary luxury marque to beat.

For the first time, Jaguar offered its top luxury car with either standard or long wheelbase. Going long gave a very significant four extra inches of legroom, in fact the longer chassis was the first to be designed, with an eye to the U.S. market. Inside, there was a choice of Luxury, Premium Luxury and Portfolio trim specifications. All of them were versions of a completely new approach to the interior which bore the leather and veneer hallmarks of Jaguar but were a contemporary take on old school elegance. A fourth level of trim, Supersport, became available to those who bought the car with its most powerful engine option.

The all-new styling of the interior was a backdrop to state-of-the-art technical innovations. There was now voice-activated control, and a virtual instrument display; a DVD hard drive, and a split screen in the transmission-mounted control panel between the front seats which allowed driver and passenger to look at different displays.

For lovers of three-letter acronyms, the safety systems built into the new car provided a host of

Above: With the long wheelbase Jaguar XJL customers could choose Luxury, Premium Luxury or Portfolio levels of trim.

examples: ABS (Anti-Lock Braking System), EBA (Emergency Brake Assist), CBC (Cornering Brake Control), EBD (Electronic Brakeforce Distribution), EBP (Electronic Brake Pre-Fill), USC (Understeer Control), ETC (Electronic Traction Control), DSC (Dynamic Stability Control), ACC (Adaptive Cruise Control) and EDTC (Engine Drag Torque Control) made up a comprehensive set of safeguards and stabilising influences. The XJ was also supplied with a bonnet which, in the event of a pedestrian accident, automatically raised itself (and any unfortunate pedestrian landing on it) clear of any sharp or hard objects beneath it – the engine block for example. This life-saving device was first seen on the XK.

Production of the XJ, originally scheduled to begin in September 2009, was delayed until March the following year, and the first cars arrived in showrooms in May. Three months later, and appropriately at the Moscow Motor Show, Jaguar unveiled a further version of the XJ, the XJ Sentinel. This was an armoured model of the long wheelbase car, with the 504bhp version of the supercharged 5.0-litre engine and run-on-flat tyres. It was capable of withstanding explosions equivalent to 15kg of TNT and with bulletproof glass strong enough to stop the hardened steel, full metal jacket of an assassin's pointed bullet. By then the Sentinel already had one happy owner – British Prime Minister David Cameron had been using one as the official car of office since May. Cameron swapped his a year later for an upgraded Sentinel, now with bomb-proof doors and armour-plating below the floorpan.

As if to allow the driving public to get used to a car which broke so many links with Jaguar's past, the production XJ has seen only relatively minor changes to its appearance and specification since its introduction. In 2011 a number of new cosmetic and comfort options were introduced. Front seats were already electronically controlled, but now a rear seat option included reclining backs and a built-in massage function.

Under the influence of its new Indian owners TATA, Jaguar turned its attention to the expanding luxury car market in China in 2012. At the Beijing Motor Show that year Jaguar presented its new XJ Ultimate, a long wheelbase model with the supercharged V8 engine linked to an eight-speed automatic transmission.

China was also chosen for the launch of a 3.0-litre V6 petrol engine version of the XJ. The engine was originally developed for the F-TYPE, and the XJ saw it linked for the first time to a four-wheel drive option.

JAGUAR XJ

ENGINE V8

CAPACITY 5000cc

BORE X STROKE 92.5 x 93mm

COMPRESSION RATIO 11.5:1 to 9.5:1

POWER 503bhp

VALVE GEAR twin overhead camshafts

FUEL SYSTEM fuel injection

TRANSMISSION 6-speed automatic

FRONT SUSPENSION independent unequal-length wishbones, coil springs

REAR SUSPENSION independent unequal-length wishbones, coil springs

BRAKES 4-wheel ventilated disc

WHEELS alloy

WEIGHT 3870 lbs (standard swb) (1755 kg)

MAXIMUM SPEED 155mph (249kmh)

PRODUCTION 2010-

Jaguar further courted China with a Hong-Kong only edition, the XJ Portfolio Prestige – which came with a 5.0-litre petrol engine, enhanced rear seat comfort option and a high-gloss oak veneer interior.

The eight-speed gearbox was rolled out for all models, and later in 2012 a supercharged version of the 3.0-litre petrol engine, and a new turbocharged i4 2.0-litre petrol engine joined the list of powertrain options. Then, at last, in 2013, the XJR arrived. Motoring journalists were delighted with the new R car, without which the XJ spectrum had felt incomplete. It was, as one reviewer described it, the flagship for the flagship range.

Not to be confused with the sports car of the late 1980s and early 1990s, the new XJR was revealed in the U.S. at the New York Auto Show, and in the UK at the Goodwood Festival of Speed, scene of so many Jaguar celebrations and innovations. It replaced the sportier iteration of the XJ, the XJ Supersport, which was discontinued with the XJR's arrival. The new model was powered, of course, by the supercharged 5.0-litre V8, with bespoke tuning which now delivered 542bhp that could deliver 0-60mph in 3.9 seconds. As with the earlier Supersport, the engine limiter was extended to 174mph, and the springs and dampers were tightened up by 30% to cope.

The forward air intakes were now outlined in chrome, and aspiration was assisted by new bonnet louvre vents. Handling was enhanced by a new front splitter, a more pronounced sill line for better aerodynamics, a rear spoiler and an electronic active differential system. Two pairs of round-section tailpipes replaced the twin flattened pipes of the XJ. It came in long and short form, with 20" alloy wheels wrapped in specially made Pirelli low-profile tyres. Inside, there were racy bucket seats in front, and aniline (naturally textured) leather trim was complimented with contrasting stitching and a choice of veneers.

Jaguar's success in the expanding markets of China and elsewhere in Asia encouraged TATA to move some production of the XF and XJ to its base in Pune, India in 2014. Now components manufactured in the UK were shipped to Pune as CKD (Complete Knockdown) kits for assembly there. The resulting vehicles were eligible for lower import duty under Indian tax law and could therefore be sold more cheaply on the sub-continent.

Below: The 3.0-litre Jaguar XJL. The 'leaper' symbol on the bootlid has become noticeably bigger over the years.

Above: The high-performance XJR-Sport is almost a match for Jaguar's top-end sports cars.

In 2016 the entire XJ range was given a significant facelift. The restyled front fascia contained a narrower, taller grille. The eye-like new lighting units were narrowed and placed higher on the wing; and like the corners of a pursed mouth the lower air intakes were narrower too. Between the latter two, large uninterrupted areas of body colour gave the whole front aspect a more imposing, aggressive aspect. The front wings were rounder, the door sills more pronounced and the rear bumper rounded off with the exhaust pipes emerging from within it. New J-shaped rear lighting units sat partly within the boot lid.

The entry-level vehicle was now called the R-Sport, with progressively higher specs available in the Portfolio, Supercharged and XJR models. The Portfolio was only offered with the longer wheelbase, the R-Sport only with the shorter.

The focus in the R-Sport was now on driver assistance. Standard equipment now included parking sensors front and back, a rear-view camera, and blindspot monitors. The long wheelbase-only Portfolio variation was distinguished chiefly in matters of trim both outside and in. It was possible to adjust the front seats, for example, in 14 ways instead of the R-Sport's mere eight. Comfort in the rear was improved with sunscreens to the side and back (the former manual, the latter electric), and new LED reading lights for those backseat drivers. Upholstery was upgraded throughout. The Supercharged XJ built on the refinements of the Portfolio with performance enhancements: a supercharged V8 engine, and an active rear differential, with tighter suspension and brakes, to handle it.

The XJR, top of the range, was powered by the 5.0-litre supercharged V8 550PS engine with an active exhaust system and suspension tuned to match its sporty ride.

The interior trim was unique to the XJR, and the exterior paint options included colours only available at extra cost on models further down the range. There were some nice touches, including a heated steering wheel; but the XJR also lost out on some of the options available to other XJs. Bucket seats, for example, meant no in-seat massage function; nor was the enhanced rear seating package available. And only the lesser of the XJ's two Meridian audio systems was fitted in the XJR. But perhaps high fidelity was less of a priority in a car that could go from 0 to 60mph in 4.4 seconds...

JAGUAR C-X75

The unveiling of a new Jaguar concept is always a big deal. But the presentation at the 2010 Paris Motor Show of the company's prototype two-door racing saloon was worthy of special attention. It was Jaguar's first electric car, an ambitious concept conceived, as the name suggests, to celebrate the 75th anniversary of the founding of the company.

In fact the C-X75 was a hybrid-electric car. Its automotive power came entirely from electric motors, fed by a lithium-ion battery. But the battery needed no lengthy stay at a roadside recharging station, it could be recharged while underway, by a pair of miniature gas turbines which were themselves driven by a variety of carbon-based fuels. On battery alone, the car had a limit of 68 miles or six hours; with the turbines activated, Jaguar claimed a range of over 550 miles.

This was no suburban run-around or electric 'roller skate'. The range alone confirmed that; but the performance was everything that would be expected from a Jaguar. The C-X75 could go from 0 to 60mph in an incredible 3.4 seconds, and its top speed tipped over the magic 200 mark at 205mph. And unlike some concepts, notably the XK120, the model that was presented was drivable and already road-tested by the time it was revealed in Paris.

This concept was very much a testbed for new ideas. Design director Ian Callum described the front face as "providing a clear confirmation of our next generation of Jaguars", while the rear (appropriately) looked back, in its seamless blend of back window and bootspace, to the XJ13. That was a 1960s one-off concept styled by aerodynamicist Malcolm Sayer, which Callum cited at the C-X75 debut as "arguably the most beautiful Jaguar ever made" – high praise from one genius to another.

The wide grille and the brake cooling ducts were part of an active aeration system, which opened automatically only when required. The whole body was aerodynamically engineered to perfection – behind the rear wheels, for example, so-called "vertical control" surfaces redirected the different airflow patterns which occurred at higher speeds, harnessing them for greater stability. The carbon fibre rear diffuser was also active,

responding to different speeds by adjusting itself to different heights and angles.

Without the bulk of a traditional engine block under the bonnet, the styling lines of the aluminium body could be lower and sleeker than ever – the highest point on the roof was less than four feet above the ground. It rode on hand-cut Pirelli tyres wrapped around wheels that could have been used for the much larger XJ range – 21" at the front, 22" at the rear. And the wheels were at the heart of this innovative vehicle. Each one was driven by a separate, independent 145kW motor. Sophisticated torque vectoring gave the potential, in theory at least, for exceptionally precise individual traction control across all speeds.

The motors were powered by a set of lithium-ion batteries which occupied the space under the bonnet where a traditional engine would have been installed. The weight savings of electric power were immediately obvious. The battery unit weighed in at 200kg, the four motors at 50 kg each, far lighter than any

JAGUAR C-X75

ENGINE 4 electric motors

CAPACITY 145kW

COMPRESSION RATIO n/a

POWER 780bhp

FUEL SYSTEM lithium-ion batteries

TRANSMISSION 7-speed automated with single clutch

FRONT SUSPENSION lever-armed double-wishbones

REAR SUSPENSION lever-armed double-wishbones

BRAKES 4-wheel ventilated carbon ceramic disc

WHEELS alloy

WEIGHT 3870 lbs (1350 kg)

MAXIMUM SPEED 205mph (330kmh)

PRODUCTION 13, 2010-2015

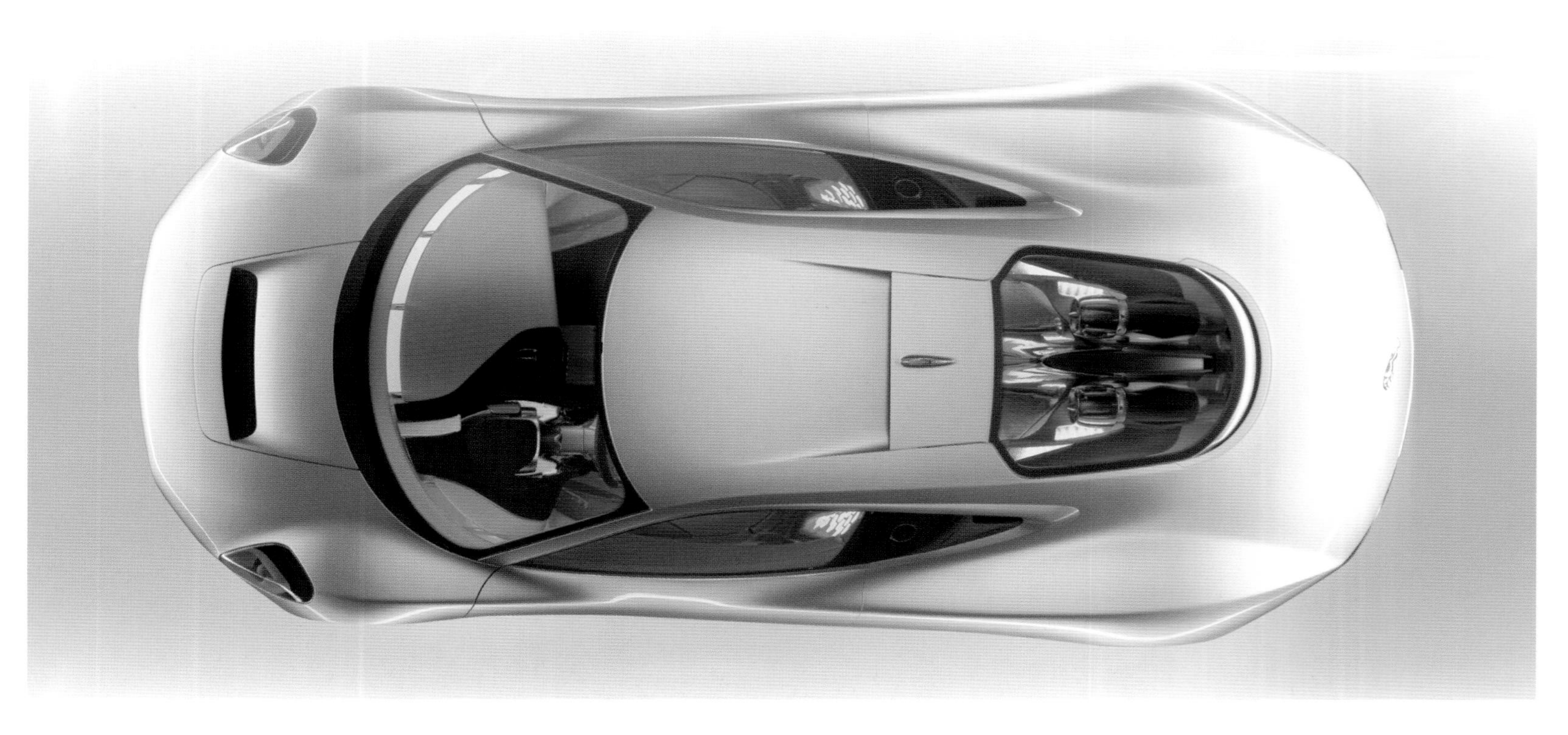

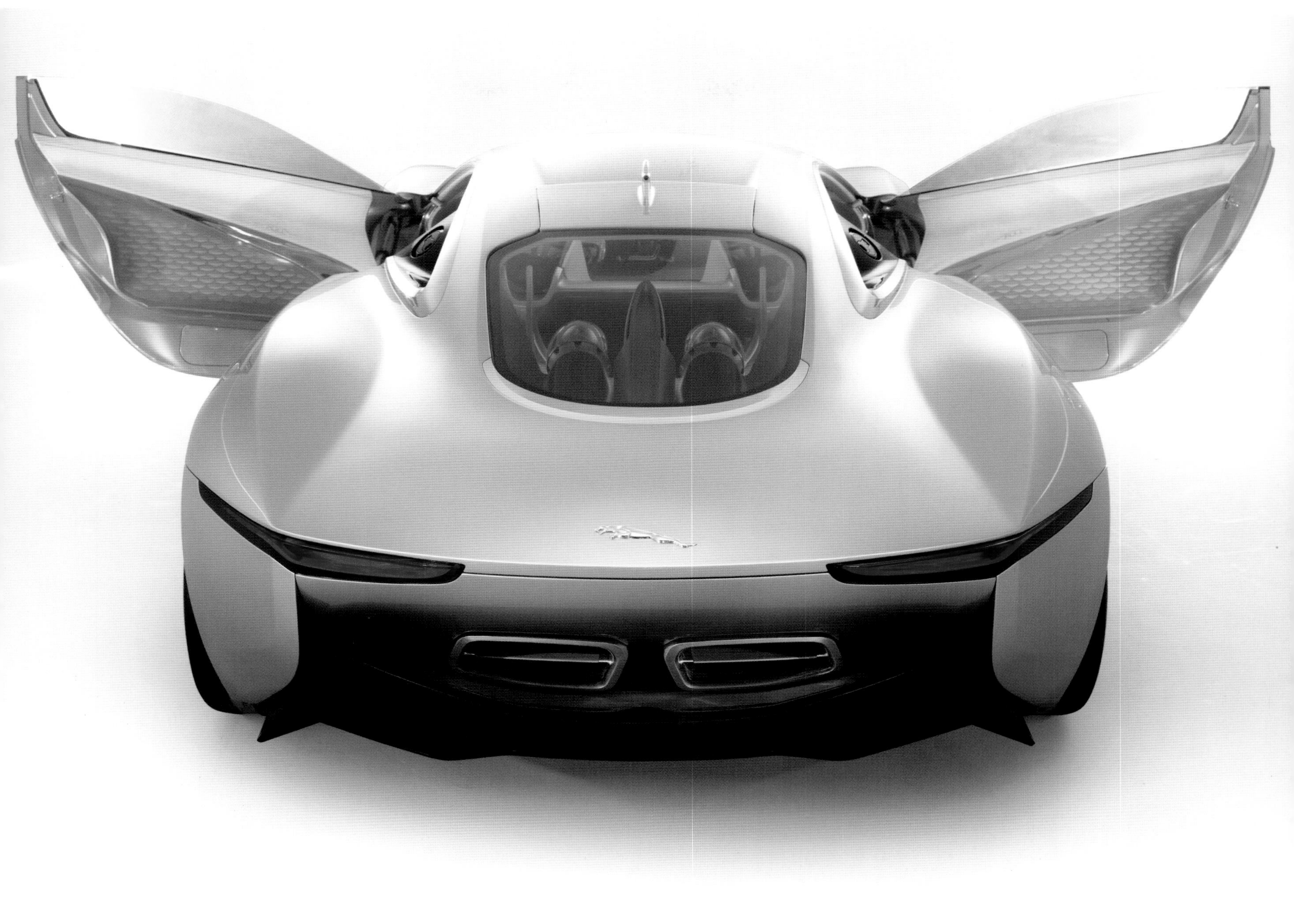

traditional combination of engine, transmission and four-wheel drive system.

In the rear, and visible through that XJ13-inspired rear window, sat the two micro-turbines (only 550mm long and 35kg each in weight) which made such a difference to the range of the C-X75. Manufactured by Bladon Jets, they sucked in up to 35,000 litres of air per minute, directed aerodynamically into the intakes on the flanks by canals pressed into the doors. The turbines compressed the air which was then burned to drive the switched-reluctance generators, which in turn charged the batteries.

Elsewhere in the rear there was a 60-litre fuel tank. The turbines might compress air but they didn't run on it – the tank could be filled with LPG, natural gas, diesel or a variety of biofuels. The car's emissions were low – only 28 grams per kilometre with the turbines running, and none at all on battery only.

Innovation didn't stop at the car's profile and engineering. Inside, the styling matched the sense of lightness which was implied by the exterior low, clean lines. Ambient turquoise lighting around the dash and in the doors defined the space, built around two low racing seats.

The seats were fixed in position – it was everything else that adjusted around them to suit the comfort and reach of the driver. The pedals, the instrument binnacle, the steering column – all of them could be moved nearer or further away, more in the manner of a jet fighter's controls than the interior of a car. But then, Jaguar had always drawn on the aviation industry for everything from its aluminium monocoque bodies (present again in the C-X75) to the very name of the company, which founder William Lyons first heard as the name of a World War One biplane's engine.

Multifunction gimbals showed road speed, power

Opposite page: The Jaguar design sketch shows the mid-1960s Le Mans car, the XJ13, at the top morphing into the C-X75 at the bottom.

Above: Jaguar loaned filmmakers the C-X75 for the villain to drive in the 2015 James Bond film *Spectre*. It took a starring role in a car chase sequence and, as at Le Mans in the 1950s, it was a case of an Aston Martin following a Jaguar.

output and turbine speed, and a separate display, described as the car's co-pilot, supported the driver with secondary information. A clock designed by Bremont was able to wind itself up simply by using the accelerating and braking motion of the car. The audio system was by Bowers and Wilkins, and the doors and rear bulkhead were literally lined with sound, in the form of a honeycomb speaker design.

The C-X75 was welcome evidence on Jaguar's 75th birthday that the company was as committed as it had always been to innovative engineering. At its launch, Jaguar Land Rover CEO Ralph Speth said that Jaguar would always continue to build beautiful, fast cars. Presented with a prime example in the form of the C-X75, motoring journalists were quick to speculate, and to hope, that it might move into production in some form.

There is always anticipation and speculation around any newly unveiled concept, and Jaguar were quick to dampen expectations by insisting that it was a one-off, an ideas machine. Five months later, at the Geneva Motor Show in March 2011, Jaguar announced a £5 billion programme of investment towards launching 40 "significant new products" by the same time in 2016. On any 40-strong list, enthusiasts felt, the C-X75 must feature quite high. And in May 2011, Jaguar announced a joint project with Formula One team Williams to build a limited edition of 250 C-X75s between 2013 and 2015.

It was not to be a straightforward reproduction of the prototype. There would now be a plug-in option for recharging, and the production chassis was to be of carbon fibre. The roofline was a little wider, the rear window a little larger. The four electric motors would

be reduced to two, one per axle. And the turbines were to be replaced by a 1.6-litre turbo supercharged petrol engine performing the same function, mounted low for stability and in order to preserve the low profile of the concept.

In early tests of the new version of the C-X75 emissions had crept up slightly, to 89g/km, but the changes had made no other concession to the car's performance. The automotive power delivered just under 900bhp and could still exceed 200mph. The acceleration was literally breathtaking, 0 to 60mph in less than three seconds, 0-100 in fewer than six.

Even with cost-saving measures for production and practicality, the petrol-electric version of the hybrid car was expected to cost any new owner somewhere between £700,000 and £1 million. Although there were still plenty of wealthy car collectors with that sort of budget, the world was taking much longer than anticipated to emerge from economic recession; and in December 2012 Jaguar reluctantly took the decision to cancel the proposed production of the 250 cars.

By then, five new prototypes had been built, and Jaguar continued to use them for research and development. The sophisticated aerodynamics of the car would certainly influence future Jaguar design, and the hybrid technology is expected to reappear, for example, by giving a three-cylinder engine the power of a straight-six. The high-pressure supercharger research is likely to be deployed on future four-cylinder Jaguars.

Above: Although the C-X75 didn't make it into production, there was still a Jaguar future for the radiator grille.

Having had their dreams of a production car dashed, Jaguar fans allowed themselves to hope again when a villain in a C-X75 was pursued through the streets of Rome by James Bond (in an Aston Martin DB10) in the 2015 Bond movie *Spectre*. Surely, they reasoned, this was a clever new marketing ploy?

Jaguar had lent the filmmakers seven copies of the car, but although they looked the part, the cars were mechanically different. Under the covers they were constructed around a spaceframe built for stuntwork to World Rally Championship standards, and the power came from a relatively conventional 542bhp supercharged 5.0-litre V8 engine. Jaguar's John Edwards, head of Special Vehicle Operations, insisted that although the film was a welcome chance to show off the production car that never was, its appearance did not indicate a change in strategy.

JAGUAR XKR-S GT

It's a measure of the reinvigoration of Jaguar under TATA's ownership that in 2013 it announced the production of a GT version of the XKR-S, the last in the line of XK cars which began in 2005. As the company settled into its role as a luxury executive carmaker, its sporting days seemed to have been consigned to history. And certainly the development costs of the new XKR-S GT ruled out any financial return from sales of the limited edition – just 30 cars, all left-hand drive for the North American market (all but five for the U.S.), and on offer for $175,000.

Jaguar took the existing XKR-S coupé and added a number of aerodynamic components – a new carbon fibre front splitter, dive planes on the front wings ahead of the XKR-S's front wheel air intakes, enlarged bonnet louvres, wheel arch extensions, a raised rear diffuser and carbon fibre spoiler. Like all XKs, its monocoque construction was all-aluminium, but now with an added aluminium undertray to improve airflow beneath the low-slung body.

Everything was designed to increase the downforce necessary to ease the handling of the supercharged V8, which was tuned like the XKR-S to give 542bhp of output. Much was made of the quadruple exhausts, described by one reviewer as 'cartoonishly loud' and by Jaguar's own marketing machine as able to 'enunciate the car's aural character'. This was a vehicle designed to frighten pedestrians.

The drive train was, perhaps surprisingly, through the XKR-S's ZF six-speed automatic transmission (an eight-speed version was already available in other Jaguars) and benefited from Jaguar's Active Electronic Differential. But the GT's electronic limiter was re-set to allow a top speed of 186mph, and an initial acceleration from standing start to 60mph which easily matched the XKR-S's 3.9 seconds.

To rein that energy in, the new car dispensed with the old brakes of the XKR-S. They were replaced by carbon ceramic units, the first time these had been used in a Jaguar production car. They were automatically pre-filled and pressurised as the driver's foot lifted from the accelerator, ready for an immediate response to the lightest touch of the brake pedal. The new discs, measured 15" in the rear wheels and 15.7" in the front, and were clamped by four- and six-piston calipers respectively.

The suspension, too, was given a makeover. The adaptive damping system could now be adjusted for height, the bushings were revised and the camber was increased. Tracking at the front was widened by two inches, although unchanged at the rear. The car rode on 20" forged aluminium wheels in gloss black with bespoke Pirelli low-profile tyres, and the steering was more responsive thanks to a quicker ratio.

There was no choice in the visual department. The new GT came in white only, with two black racing strips along its bonnet. Inside, everything from the 16-way-adjustable bucket seats to the steering wheel was charcoal grey, with red piping on the upholstery and suede cloth to the seat backs, the headlining and the steering wheel. The rear seating was dispensed with altogether to make room for a comprehensive roll cage. This was a car where all the attention was on performance, none on looks. As chief designer Ian Callum remarked, it was "raw, focused and devastatingly quick."

The GT was presented to the world at the 2013 New York International Auto Show, a surprise companion to the XJR which was unveiled for the first time at the same show. It was a double statement of Jaguar's renewed sporting energies, and the XKR-S GT appeared to the delight of Jaguar enthusiasts, except perhaps those in the UK who were to be denied the chance to own one.

Following its New York debut, it became clear that the GT was something of a testbed for Jaguar's ETO (Engineered To Order) division, which conducted closed trials on road and track in Germany and the UK. The dampers, springs and uprights of the improved suspension, for example, were expected to become standard for future sporting Jaguars including the F-TYPE. The GT also sported a new rear axle. The carbon ceramic brakes overcame a criticism of the

JAGUAR XKR-S GT

ENGINE supercharged V8

CAPACITY 5000cc

BORE X STROKE 92.5 x 93mm

COMPRESSION RATIO 9.5:1

POWER 542bhp

VALVE GEAR twin overhead camshafts

FUEL SYSTEM fuel injection

TRANSMISSION 6-speed automatic

FRONT SUSPENSION independent unequal-length wishbones, coil springs, adaptive dampers

REAR SUSPENSION independent unequal-length wishbones, coil springs, adaptive dampers

BRAKES 4-wheel carbon ceramic discs

WHEELS forged aluminium

WEIGHT 3781 lbs (1713 kg)

MAXIMUM SPEED 186mph (299kmh)

PRODUCTION 40, 2013-2014

Left: At launch, the XKR-S GT was only available in white and only available in left-hand drive.

P ZERO

stopping power of the original XKR-S.

All the innovations and fine-tuning paid off. Later in the year, the XKR-S GT recorded a lap time on the Nürburgring's Nordschleife track of 7min 40sec, the fastest time ever recorded there by a road-legal Jaguar and a speed to challenge many supercars. The record was a clear message to its rivals – Jaguar was interested in speed again.

It came as no surprise when, at the end of the month, Jaguar announced that it would be building a further ten XKR-S GTs – and this time they would be right-hand drive, retailing in the UK for £135,000 apiece.

One of the great successes of the ETO engineers was to produce a car which drove as well on the road as on the track. The suspension was 68% firmer in front and 25% behind and the XKR-S GT was stiffer at legal speeds than the XKR-S. But it was noticeably kinder to its occupants than the XFR-S, which was almost too firm for everyday driving. At 23mpg it was relatively fuel-efficient for its class too.

Jaguar must have been very satisfied with the technology brought to bear on the XKR-S GT. It delivered a road-legal sporting coupé to rival many supercars; and a track-worthy car to honour the marque's sporting history. It was never expected to cover its costs. But as an investment in future technology, and as a poster boy for the entire Jaguar range, it certainly paid dividends. In the twelve months following its launch in New York, Jaguar sales rose by 30%.

Above: The 'R' in all Jaguar models stands for 'Racing', but when an 'S', a 'GT' and a large rear wing are added, this is clearly not a car for going to the shops.

Left: Prior to the launch of the XKR-S GT, the XKR had raced successfully with a wider, straight rear wing. Here seen at Mosport Raceway in Canada.

JAGUAR F-TYPE

At the end of the first decade of the 21st century Jaguar's confidence was as high as it had ever been. New owner TATA made buses, trucks and small, affordable, mass-market cars. Its portfolio included no vehicles in any of the markets occupied by Jaguar, unlike Jaguar's previous parent Ford, and there was far less pressure on Jaguar to compromise or share technology. Ownership by an Indian company had opened up new Asian markets and sales were booming.

In 2010 the Jaguar range included a luxury car (the XJ), an executive car (the XF) and a grand tourer (the XK). All of them had been launched after the millennium; all to some degree boasted radical designs which made a break with Jaguar's illustrious but sometimes overwhelming past. Jaguar was even involved with racing again, in the form of the XKR GT. There was just one thing missing from the range: a genuine Jaguar sports car.

Production of the iconic E-type had ceased in 1975, and all subsequent models with claims to the crown had in reality been to a greater or lesser degree grand tourers rather than sports cars. Enter, in 2013, the F-TYPE, a genuine successor in name, in body and in performance. With such a strong portfolio of 21st century designs in the Jaguar range, the company was confident enough in its future to revisit its E-type past without accusations of clinging on to former glories.

The F-TYPE had a more immediate lineage than the

E-type. It was derived in large measure from a concept car, the C-X15, which was presented at the Frankfurt Motor Show in 2011. Although it adopted a slightly wider version of the classic horizontal E-type oval grille and its narrow wraparound rear lighting unit, the C-X15 skipped straight back to the E-type for its side-hinged coupé rear window. All three features – grille, lighting and window – would resurface on the F-TYPE.

Less than fifteen feet in length, the C-X15 was the shortest prototype which Jaguar had produced since it launched the very sporting XK120 in 1954, nearly 60 years earlier. The signs were there for seasoned Jaguar watchers to see that the company was working towards something small and aggressive.

When it appeared the F-TYPE used a shortened, all-aluminium form of the XK's chassis. Just one inch longer than the C-X15, it was strictly a two-seater and, at launch, only available as a convertible. That launch, after a preview at the London edition of the Sundance Film Festival in January 2012, came at the 2012 Paris Motor Show. A formal British debut came at Goodwood in July 2013, the same event at which Jaguar unveiled the XKR-S GT. By the end of the year Jaguar was presenting a companion for the convertible, the fixed-head F-TYPE coupé.

There was in truth something of the E-type about the F-TYPE, although the more streamlined angle of

Below: The C-X15 made its debut at the Frankfurt Motor Show in 2011. The F-TYPE convertible would follow at the Paris Motor Show in 2012.

JAGUAR F-TYPE

ENGINE V8

CAPACITY 5000cc

BORE X STROKE 92.5 x 93mm

COMPRESSION RATIO 9.5:1

POWER 488bhp

VALVE GEAR twin overhead camshafts

FUEL SYSTEM fuel injection

TRANSMISSION 8-speed automatic

FRONT SUSPENSION independent wishbones, coil springs

REAR SUSPENSION independent unequal-length wishbones, coil springs

BRAKES 4-wheel ventilated discs

WHEELS alloy

WEIGHT 3671 lbs (1665 kg)

MAXIMUM SPEED 186mph (299kmh)

PRODUCTION 2013-

Above: Reviewing the F-TYPE coupé, *Top Gear* magazine described it as 'arguably more fun than a Porsche 911'.

the windscreen made the cockpit less of a bubble than the old classic. The grille was a nod to the past, as were the headlights set into the wide front wings. Between them the power bulge of the bonnet swooped down to the growler badge set not in the middle but at the top of the mesh. Additional air intakes flanked the radiator, assisted by the trademark horizontal louvres forward of both doors. Well-defined sills drew the eye to the high rear wings, more pronounced on the convertible in the absence of the coupé's rear hatch.

The convertible's roof was of fabric and could close in 12 seconds at the touch of a button, even with the car on the move at up to 30mph, stowing itself behind the seats and becoming its own tonneau cover. It incorporated a layer of fabric supplied by the Thinsulate all-weather clothing manufacturer which, Jaguar claimed, was almost as effective as a solid roof in terms of thermal and aural insulation. Noise and vibration were potential problems in such a small, powerful car, and Jaguar's engineering solutions included a twin-layer bulkhead between the occupants and the engine bay, special engine mountings, and an underbody tray.

Access to the car was by touch – or rather, by concealed door handles which popped out of the bodywork when touched. Inside, the leather trim was immaculate as ever, but more understated than the luxury cars of the line. The hard details, knob and switch finials for example, were in aluminium rather than walnut. Without rear seats, the front legroom was relatively generous, and the seating itself was low on the floor. Sports seats however were an option, not as standard.

With very little overhang beyond the rear axle, the boot was a rather small affair, especially in the convertible which did not have the extra volume created by the coupé's hatch. Behind the folded soft-top there was just 7.1 cubic feet (201 litres) of storage space.

Navigation along the way was assisted by a central sat-nav touchscreen from which the audio system was also controlled. Audio was once again by Meridian, with a choice of a 380w ten-speaker installation or a 770w twelve-speaker upgrade.

The F-TYPE was powered by one of three supercharged petrol-driven powertrains, all via a ZF 8HP eight-speed automatic transmission. Jaguar's AJ126 engine was a 3.0-litre V6 delivering to the entry-level F-TYPE, 335bhp and a maximum speed of 161mph, with 0-60mph acceleration of 5.1 seconds. The F-TYPE S used the same engine which was retuned to give 375bhp and reach 171mph.

Top: The door handles of the F-TYPE revealed themselves when touched.

Above: The F-TYPE gained its own distinctive Power Vent.

At the top of the range, the F-TYPE V8S convertible housed (as its name suggests) an AJ133 5.0-litre V8 engine putting out 488bhp to get to 60mph in 3.9 seconds. When the coupé arrived in showrooms in 2014, its V8 incarnation, the F-TYPE R, was further tuned to deliver an extra 54bhp, with All-Wheel Drive. The engines weren't the only things that got bigger as you moved up the range: the entry-level F-TYPE's alloy wheels were 18", the F-TYPE S's 19", and those of the V8 model a full 20". All were fitted with bespoke Pirelli low-profile tyres.

In 2015, the much-anticipated V6 machines were offered with a new choice of transmission: the eight-speed automatic, or ZF's 6HP six-speed manual gearbox , making it the first Jaguar sports car to feature a manual gearbox since the XJ-S. Included as standard and mounted on the steering column of all automatic F-TYPEs, Jaguar's paddle-shift allowed automatic drivers to take manual control of gear changes if they

Left: The fabric roof to the convertible could be closed in 12 seconds at speeds up to 30mph.

Below: The F-TYPE S convertible has a V6 engine and two tailpipes. The F-TYPE R is equipped with a V8 and quad tailpipes.

preferred; and a new stop-start button shut the engine off in stationary traffic, saving (Jaguar claimed) up to 5% of fuel costs.

Suspension was sophisticated across the F-TYPE range, with adaptive dampers linked to twin wishbones front and back. Precise responses were programmable by the driver, who had a choice of 25 different suspension modes tailored to 25 different driving styles or road conditions.

It was expected that, however you drove, you would drive the F-TYPE to its limits. For drivers who wanted everyone else to know they were driving to the limits, the S and V8S versions came with an active exhaust system. It opened special valves in the tailpipes to amplify the roar at engine speeds of over 3000rpm.

The F-TYPE was promoted with a sophisticated campaign of celebrity endorsement. American pop singer Lana Del Rey was enlisted for its Paris launch in 2012; Former Real Madrid and Chelsea Football Club manager Jose Mourinho was made an 'ambassador' for Jaguar UK and took possession of the first F-TYPE Coupé in 2014; while a very British soccer giant, David Beckham was cast in the role of ambassador for the coupé's launch in China. Beckham's twin careers as footballer and model for male grooming products were obvious choices for association with the style-conscious, sporting F-TYPE.

Below: The convertible came before the coupé due to its bigger sales potential in the key American market.

OY14 HGL

JAGUAR F-TYPE PROJECT 7

With one eye on the AMG branding at the top end of the Mercedes range, the Jaguar Land Rover group launched their Special Vehicles Operation in 2015. The new 20,000m² facility on the site of the old Peugeot factory in Coventry – complete with VIP reception centre – was a far remove from the old Jaguar Competitions Department at Browns Lane.

There were also echoes of McLaren's Special Operations division at Woking and a Formula One connection in that the man Jaguar put in charge of the technical side, Paul Newsome, came from the Williams Advanced Engineering group.

There have been many connections between the two organisations in the past. When he was at Williams, Newsome helped develop the Jaguar C-X75 hybrid supercar, and going forward, Williams Engineering supply the batteries used in the Formula E single-seater racing series. Jaguar entered the third series of FE in Autumn 2016 with a team run by Williams. (Much to the consternation of other teams who thought Jaguar would get privileged advice into the deployment of the energy source!)

Newsome's first task was making the beautiful F-TYPE Project 7 sportscar a roadgoing reality. With a long history of producing concept cars that never quite made it into production, the company had originally intended it as a celebration of Jaguar's seven Le Mans victories, lodging the '7' prominently in the model's name. It was styled by Ian Callum with a nod to the Jaguar D-type with its rollover hoop and rear faring behind the driver's head, and the new car had unmistakeable Jaguar lines.

The company took it to the 2013 Goodwood Festival of Speed and the Pebble Beach Concours d'Elegance and enquiries were brisk. A decision was taken to make a limited run of 250 cars that would be hand-built by Jaguar Land Rover's new Special Vehicles Operations facility. The car was going to be a lot harder-edged than simply restyling an F-TYPE R. Jaguar's short-screen speedster would be the most powerful production car

DUNLOP
JAGUAR

that the company had yet assembled (before the arrival of the 2017 Jaguar F-TYPE SVR). The 5.0-litre V8 supercharged engine was modified to produce 567bhp, 25bhp more than an F-TYPE R Coupé.

Turning what was a concept car into a production car was not such a routine engineering proposition. A lot of work was undertaken in the wind tunnel and the car took to the test track to be re-engineered and fine-tuned before Newsome signed it off. "In a high-power, short-wheelbase car like this, especially if it has a limited-slip diff and stiffened front end, engineers have to cope with two effects: a tendency towards too much understeer in corners and a loss of traction on the inside rear wheel," he said. "We increased front negative camber from 0.5° to 1.5°, to encourage the front wheels to dig in, and we used rear torque vectoring – differential braking of the rear wheels – to make the car turn easily. It's surprising how much yaw you can achieve with relatively little braking, although

the challenge is to make it feel natural."

To justify its existence – apart from the cachet that a limited production run would bring – the Project 7 needed to be demonstrably more agile than the F-TYPE convertible. SVO engineers were able to rebalance the car's rear-biased aerodynamic downforce by fitting blade-like side skirts and a large carbon fibre front splitter, while slightly reducing the drag from the rear wing.

A significant 45kg weight reduction contributed to a 0-60mph sprint time of 3.8 seconds and an electronically limited top speed of 186mph, as with other F-TYPEs. The Project 7 was also given unique race-style springs and dampers, track-focused suspension geometry, enhanced steering response and standard carbon ceramic brakes (with 398mm six-pot front discs).

Engineers also moved the settings for Sport and Standard suspension further apart, to distinguish between its stiffened track-day use and, say, a jaunt to Le Mans that might include an encounter with pot-holes or traffic calming measures. With that in mind, to impress the nearest petrolhead, a special console switch opened the exhaust butterflies and allowed the car to emit a raucous, growly-bark exhaust note, with a menacing crackle on the overrun. Tested around the Nürburgring it produced a lap time of 7min 35sec, four seconds quicker than the previous Jaguar record holder, the F-TYPE R Coupé, making it worthy of its Le Mans-inspired name.

Previous page: The Jaguar F-TYPE Project 7 at its spiritual home, Le Mans.

Left: Jaguar estimated that the serious car collectors who ordered the Project 7 would do no more than 2,000 to 3,000 miles a year in their cars, whereas after just two years on sale, F-TYPEs were appearing on the second-hand market with 30,000- 40,000 miles on the clock.

JAGUAR F-TYPE SVR

Whereas the Jaguar F-TYPE Project 7 had been a limited production run model – and swiftly sold its 250 capacity – the F-TYPE SVR was the first series production Jaguar to be developed by the Special Vehicle Operations division. The production run would continue to meet whatever demand there was for what would be the fastest Jaguar in the range and a serious threat to the Porsche 911 GT3.

The Jaguar Land Rover group had started SV badging with the Range Rover Sport, which was transformed into the Range Rover Sport SVR, a vehicle well-received across the motoring press and a sales success from the moment it arrived in the showroom.

Jaguar were next in line for attention from the Special Vehicles Operation with their F-TYPE R Coupé and Convertible gaining significant SV upgrades to become the F-TYPE SVR. Whereas the highly sporty F-TYPE Project 7 was an edgy, track-day gem eager to get off the leash and leave rubber on tarmac, the SVR was an all-weather 200mph supercar that could be driven every day of the week.

The Jaguar F-TYPE SVR for 2017 was launched at the Geneva Motor Show in March 2016 intended as the first of what could become many SV versions extending down the range of Jaguar models. The 'R' represents Racing and with that ethos in mind Jaguar have stated that an SVR car will need to be '…lighter, more aerodynamically efficient and more powerful than any other model in its range.'

The F-TYPE SVR develops 575bhp from its 5.0-litre supercharged V8, which is 25bhp more powerful than the F-TYPE R. The SVR's aerodynamic package is an uprated version of that used by the F-TYPE R and it comes equipped with All-Wheel Drive and an eight-speed automatic gearbox.

As part of the aero package, the SVR has been given a revised rear diffuser and an active rear wing that extends when needed to add downforce and stability in cornering as the car picks up speed, however to attain the very top speed claimed for the car it needs to be lowered entirely.

Engineering lessons have been learned from the testing of the Project 7. Along with a new front end, a flat underfloor and a carbon fibre active rear wing, the chassis has been uprated, with new dampers and anti-roll bars added. Extra grip is gained through wider Pirelli P Zero tyres applied to lightweight 20" forged alloy wheels.

Fitted as standard is the (self-aggrandizing) Jaguar Super Performance braking system with large, 380mm and 376mm front and rear brake discs. The F-TYPE SVR can also be specified with a carbon ceramic matrix (CCM) braking system, with even larger 398mm and 380mm discs and six- and four-piston monobloc calipers. The brake prefill system installed in the car is designed to ensure consistent pedal feel.

Jaguar describe the new titanium exhaust as giving 'an even more purposeful, harder-edged sound' than the F-TYPE R, making it sound as fast as it looks.

In performance terms, the TYPE R was a hard act to follow, considering it could crack a top speed of 186mph. The 1705kg F-TYPE SVR coupé is 25kg lighter than the AWD F-TYPE R and lighter still if all of the carbon fibre options are taken. It can manage 0-60mph in 3.5 seconds and reach a 200mph top speed. The convertible is heavier at 1720kg and can match the 0-60mph time of the coupé but its top speed is 5mph slower. CO2 emissions figures are the same for both cars, at 269g/km respectively.

Left, Right and Below: The All-Wheel Drive F-TYPE R had already pushed the performance barriers for Jaguar, but the introduction of the SVR and its lightweight options made it the ultimate speed machine in the Jaguar range.

JAGUAR F-TYPE SVR

ENGINE V8

CAPACITY 5000cc

BORE X STROKE 92.5 x 93mm

COMPRESSION RATIO 9.5:1

POWER 575bhp

VALVE GEAR twin overhead camshafts

FUEL SYSTEM fuel injection

TRANSMISSION 8-speed automatic

FRONT SUSPENSION aluminium double wishbones

REAR SUSPENSION aluminium double wishbones

BRAKES Jaguar Super Performance Braking System, 4-wheel ventilated discs, optional Carbon Ceramic Matrix (CCM)

WHEELS Coriolis lightweight 20-inch forged alloy

WEIGHT
Coupé 3759 lbs (1705kg)
Convertible 3792lbs (1720kg)

MAXIMUM SPEED
Coupé 200mph (321 kmh)
Convertible 194mph (312 kmh)

PRODUCTION 2016-

SPECIAL VEHICLE
SVR
OPERATIONS

INDEX

First published in the United Kingdom in 2016 by
Collins & Brown
1 Gower Street
London
WC1E 6HD

An imprint of Pavilion Books Company Ltd

Distributed in the United States and Canada by
Sterling Publishing Co., Inc.
1166 Avenue of the Americas,
New York, NY 10036

ISBN 978-1-91121607-0

A CIP catalogue record for this book is available from the British Library.

10 9 8 7 6 5 4 3 2 1

Editorial content supplied by Salamander Books
Reproduction by Mission, Hong Kong
Printed and bound by 1010 Printing International Ltd, China

This book can be ordered direct from the publisher at www.pavilionbooks.com

Picture Credits:
The publishers wish to thank the following for supplying editorial photos for use in the book:
Pavilion Image Library: 2, 6 (bottom right), 7 (bottom left), 18, 19, 20 (bottom), 21, 22, 24, 26 (top), 27, 28, 29, 30-31, 32, 33, 34, 35, 37, 38, 44, 45, 48, 49 (right), 52, 53, 55, 56, 57, 60, 61, 65, 66, 67 (top), 68, 69 (bottom), 70, 71, 74, 75, 81, 83, 86, 87, 92, 94, 100, 101,102, 103, 104, 105, 106,109, 110, 112, 113, 114, 115 (top left), 115 (top right), 116, 117, 118, 120, 121, 125, 130, 131 (bottom), 159.
MagicCarPics.co.uk: 4, 6 (left), 7 (top left), 9,10, 14, 15, 16, 17, 20 (top), 26 (bottom), 36 (left), 39, 46, 47, 49 (left), 50, 51, 58, 59, 62, 63, 69 (top), 72-73, 80, 82, 82-83, 84-85, 96, 97, 98, 99, 107, 108, 111 (bottom), 119, 122-123, 126, 127, 132-133, 133, 134-135, 136, 137, 140 (top), 140 (middle), 142-143, 151, 152-153, 154, 155, 156-157, 158, 160, 162-163, 165 (top), 166-167, 188-189, 189 (bottom), 190, 191, 194-195.
Jaguar Cars: 11 (bottom), 13, 79 (bottom), 90, 91 (bottom), 95 (bottom), 115 (bottom), 131 (top), 138, 139, 141 (right), 144, 145, 146, 147, 148, 149, 150, 165 (bottom), 168-169, 170, 171, 172, 173, 174-175, 176, 177, 178-179, 180, 181, 182-183, 184, 185, 186, 187 (bottom), 189 (top), 192, 197-198, 199, 200, 201, 202-203, 204-205, 205, 206, 207.
Getty Images: 7 (top right), 8, 11 (top), 23, 40, 41, 42-43, 77 (bottom), 78, 93, 128, 161, 164, 187 (top).
Alamy: 6 (top right), 64, 67 (bottom), 79 (top), 88-89, 91 (top), 124, 193, 198, 198-199.
Bonhams Auctioneers: 76, 77 (top).
Norbert Aepli: 153.
Brian Snelson: 95 (top).
Dimitry Valburg: 12.
Cover image: Magic Car Pics (Magiccarpics.co.uk).